Nests, Birds a

New insights into natural and artificial incubation

D. C. Deeming

Brinsea Products Ltd.

2002

Nests, Birds and Incubators: New insights into natural and artificial incubation

Denis Charles Deeming*

Publisher: Brinsea Products Ltd**.
               Station Rd., Sandford, N. Somerset, BS25 5RA, United
               Kingdom.

Printer: Oxford Print Centre,
               36 Holywell Street, Oxford, OX1 3SB, United Kingdom

**www.Brinsea.co.uk
*www.deemingdc.freeserve.co.uk

ISBN 0 9544067 0 2 paperback

Preface

My concept for the book "Avian Incubation: Behaviour, Environment & Evolution" (Oxford University Press, 2002) was to bring together a broad range of papers on natural incubation in birds that had not been summarised together previously. I am proud of the end result and I hope that now the scientific community has a book on natural incubation that will be a reference text for years to come. However, whilst I would like to hope that the book is readily accessible to the scientifically minded reader, I appreciated that it would be hard-going for a lot of people interested in incubation. I was particularly pleased when the opportunity arose to present much of the same information in a way that would be easily read by non-scientists. It was also exciting to attempt a fusion between natural and artificial incubation and therefore, investigate the similarities and differences between these two seemingly divergent procedures. The end result is a unique insight into bird incubation that I hope will serve to inform those people breeding exotic bird species and relying on either natural or artificial incubation, or perhaps a combination of the two.

I would like to thank Frank Pearce and Ian Pearce of *Brinsea Products Ltd.* for the opportunity to write this book. Frank read through the text as it was written and provided many useful comments on the text and was able to spot some very obvious corrections. During the course of writing this book, Frank and I have learnt a lot about natural and artificial incubation from our own different perspectives and I hope that we are both the better for the experience. Thanks go to Ian for his editorial comments and for driving halfway across England to bring an example of the contact incubator for me to see. I would like to thank Dave Phelps for his assistance in improving a few of the illustrations and for such a good cover.

Writing a book of this type involves considerable time and effort and as a result it can tax the patience of those around you. As usual my family have been of great support – at times my 18 month-old daughter even wanted to help. Many thanks go, therefore, to my wife Roslyn and my daughters, Katherine and Emily, for putting up with me spending hours upon end staring at a computer screen or with my nose in a book.

D. C. Deeming
Lincoln, December 2002

Contents

Foreword by Frank Pearce

Dr. Charles Deeming is probably one of very few scientists across the world with the particular knowledge and understanding to be capable of writing this book. He is the author of numerous scientific papers concerning subjects as diverse as the physiology of egg turning to the management of ostriches in temperate climates. He has edited and co-authored various scientific books but has also written several books, including "Ratite Egg Incubation" and "Principles of Artificial Incubation for Game Birds", aimed at people interested in the practical aspects of incubation.

Since graduating from Bath University in 1984 his career has combined a remarkable mix of academic research and hands-on practical involvement with incubation and rearing of a range of species. I have known Charles personally since we became friends at "Incubation Research Group" conferences in England in the early 1980s. This group, now known as the "Incubation & Fertility Research Group" has been co-ordinated and contributed to by Charles many of the years since, and both he and the group have been a valuable source of insight to me.

The concept for "Nest, Birds and Incubators" arose out of a long held belief that incubation of wild or undomesticated bird eggs needed a fundamental rethink based on a close study of natural incubation, rather than sole reliance on further development of conventional poultry techniques. The huge reserves of information in the scientific literature have largely remained the preserve of the academic community and as long ago as 1988, Charles and I discussed possible collaboration on a book to cross this void to provide better information to breeders. Eventually it was the publication of "Avian Incubation: Behaviour, Environment and Evolution" in 2002, which Charles conceived and edited, that prompted me to encourage Charles to write the present book. 'Avian Incubation' is a remarkable collection of scientific papers on different aspects of natural incubation by world-renowned specialists, including Charles. As such the book broke new ground in bringing together such diverse expertise specifically on the subject of natural incubation.

"Nests, Birds and Incubators" is an attempt to set out much of this specialised scientific data on wild bird behaviour, as well as more widely known studies into poultry incubation in a readable way, and then to apply it to the design and application of incubators. The final chapter deals with a fundamentally different approach to incubator design, which may offer zoologists and conservationists a much closer approach to natural incubation for the future.

1 - Incubation – A Critical Aspect of Reproduction in Birds

Birds form a significant part of human life. Being active during daylight they are a familiar part of our daily activity and their behaviour has enraptured people for millennia. Keeping birds in captivity has long been a tradition in many human societies where their colourful plumage and songs have endeared birds to our hearts. In more recent years the biology of birds has been the subject of intense scientific study and their reproductive behaviour is of particular interest. Part of this interest comes from the exploitation of birds for their flesh, their eggs or (in the case of ostriches and other ratites) their feathers and skin. More pure scientific studies have revealed great detail about the social and reproductive lives of a vast range of bird species. Much of this information has helped in the development of artificial techniques to maintain birds in captivity and in particular, to breed them for food, profit, conservation purposes, or simply for pleasure.

In this introductory chapter I provide a brief outline of the scope of this book. This will highlight the topics of interest in each chapter. Thereafter, I feel that it would be useful to place incubation into its broader context of bird reproduction and there is a brief description of the key aspects of breeding in birds.

The scope and aims of this book

This book aims to describe incubation of bird eggs in 1) a natural situation, *i.e.* sitting on eggs in a nest, and 2) an artificial situation where a machine is used to incubate and hatch the birds. This approach has been taken in order to highlight the similarities and differences between the two systems in the hope that a better understanding of the factors affecting natural incubation will lead to improvements in the success of artificial incubation. The content of the book is based on the latest scientific understanding of these events and procedures. Although much of this information is published it is often in a form that is not very accessible to non-scientists. Here I have attempted to make these sometimes quite complex ideas accessible to enthusiastic breeders who want to learn more about the birds' behaviour as well as wanting to improve their results during artificial incubation.

The first part of the book deals with natural incubation and describes the basis of the "bird-nest incubation unit". The interaction between the bird and the nest to ensure that the egg is incubated properly is considerable; the role of the nest, egg or bird during incubation cannot be taken in isolation. The following chapter describes the variety of nest types together with functional aspects of nest structure. The next chapter describes egg formation, composition and structure. The development of the embryo from fertilisation through to hatching is described in chapter 4. The final chapter in the first part of the book aims to describe the behaviour of adult birds during incubation and how they achieve the appropriate incubation conditions for their eggs.

The second part of the book deals with techniques of artificial incubation. Chapters focus on why artificial incubation is such a valuable tool and how it has been developed over the years. A critical chapter describes the principles of how different types of incubator work and how they should be managed to get the optimum performance. Emphasis is placed on small-sized incubators typically used for small breeding operations. Another chapter deals with the broader aspects of incubation management, *e.g.* egg handling and procedures like monitoring egg weight loss. Given that artificial incubation places the control for embryonic development on the person rather than the bird, I use one chapter to provide ideas of the things that could go wrong and what to look for if you have a problem. The final chapter describes a new concept in artificial incubation that blurs the previously clear distinction between bird and machine. Development of the artificial contact incubator has been possible as people have begun to better understand why and how birds incubate their eggs.

Books on incubation for bird eggs can either be very general or quite specific in the type of bird they cater for, *e.g.* ratites or game birds. Luckily, bird eggs are pretty conservative in terms of the incubation conditions they require. This has meant that the eggs from all species can be incubated in machines. The degree of success depends heavily upon the understanding of the natural behaviour of the birds in question. For instance, the temperature employed to incubate the large eggs of ratites (*e.g.* the ostrich, the emu and the rheas) is about a degree Celsius lower than that used for poultry eggs. This is not due to a lower temperature for embryonic development in these species but rather it reflects the thermal characteristics of these large eggs. The latter retain heat more readily than smaller eggs and so the machine has to be set at lower temperature in order to ensure that there is sufficient cooling during the later stages of development. At the other extreme, successful artificial incubation of very small eggs may perhaps be compromised by the turning mechanism of the incubators.

This book cannot deal in specific detail with all of the possible species being incubated around the world. This is primarily because I intend to describe incubation in nests as well as in machines. To avoid the text getting over complicated I have concentrated on groups of birds where artificial incubation is commonly employed. These include: parrots, waterfowl, birds of prey (raptors), cranes and bustards, penguins, ornamental poultry and game birds, ratites (including kiwis), and to a lesser extent songbirds. These groups serve to cover the range of bird reproduction: from some of the smallest to the largest eggs, from basic nest scrapes to complex nest structures, from bitterly cold to blisteringly hot climates. It also covers the range of developmental modes adopted by birds. This is the spectrum of hatchling types ranging from the bald, blind altricial young of parrots and songbirds through to the fully independent young of waterfowl and galliforms (see pp. 7-8). Therefore, examples shown in this book will generally draw on these groups of birds although typically the principles of incubation will apply to all other species.

This book is aimed primarily at the non-scientist but the nature of the topics described mean that it is hard to avoid discussing ideas and subjects that can be highly technical. Whenever possible I attempt to avoid too many technical terms and any excessive detail in the scientific study or explanation of any particular aspects of incubation. For instance, I fully appreciate that most people have no wish to know the fine details of the thermal characteristics of the egg under a bird. However, there are times when it is not possible to avoid technical terms and ideas. I deal with these in two ways. Firstly, there is a glossary of technical terms at the end of the book, which will help the read to understand some of the technical terms. Secondly, when appropriate (for example, the concept of hatchling maturity) I will describe particular aspects of the topic in self-contained boxes within the text. These provide more detail without breaking the flow of ideas being discussed in the main text. Anyone wishing to learn more about any particular topic can refer to the reading list provided.

Bird reproduction

Whole books have been written on bird reproduction and in the following section I have no intention of describing in any great detail the various strategies birds adopt when breeding. I do hope, however, that the brief description that follows gives an idea of the behavioural and physiological changes that are happening during the breeding cycle (Figure 1.1). Incubation plays a critical role in the reproductive activity of birds because it links courtship and mating with the process of rearing of the offspring.

Control and timing of reproductive behaviour

The exact timing of reproduction in birds is under hormonal control alth-

Figure 1.1. A summary of the key stages in the cycle of bird reproduction.

ough the secretion of the reproductive hormones by the gonads (*i.e.* the sex organs – the testis in males and the ovary in females) is affected by a variety of factors. Although the exact timing of breeding may be modified by "ultimate" factors, such as food availability or the weather, the hormonal process of breeding control is ultimately affected by more fundamental "proximate" factors such as day length.

In species that breed in areas where day length changes considerably with the season (Figure 1.2A), the short days of autumn stimulate the pineal gland in the brain and this controls production of gonadotrophin-releasing hormone (GnRH) from the brain (Figure 1.2B). This hormone stimulates production of other hormones (gonadotrophins) from the pituitary that promote the growth of the gonads (Figure 1.2B) and secretion of other sex hormones (oestrogen from female ovaries and testosterone from male testes) important in behaviour and physiological changes, such as egg and sperm production. The amount of GnRH produced is proportional to the amount of daylight up to a threshold of day length, above which hormone secretion is suppressed (Figure 1.2B). Therefore, the short days of winter produce low levels of GnRH that help to slowly bring the bird into breeding condition. As day length increases in the spring, GnRH secretion increases and breeding activity is maximised at the correct time for good climatic conditions and food supply for the adults and chicks. As the year progresses the long days during summer eventually cross the day length threshold cutting off GnRH secretion which inhibits sex hormones secretion and gonad size decreases (Figure 1.2B) as the birds lose their reproductive urge. As indicated above the exact timing of courtship and nesting will be modified by ultimate factors during any particular spring season. For instance, cold weather during early spring will delay nesting and egg

Figure 1.2. Diagrammatic representation of the annual changes in the A) day length, B) GnRH secretion (Solid line) and gonad size (dashed line) in a temperature climate, and C) percentage of birds in a population nesting (white columns), incubating (light grey columns), feeding chicks (dark grey columns) and moulting (black columns).

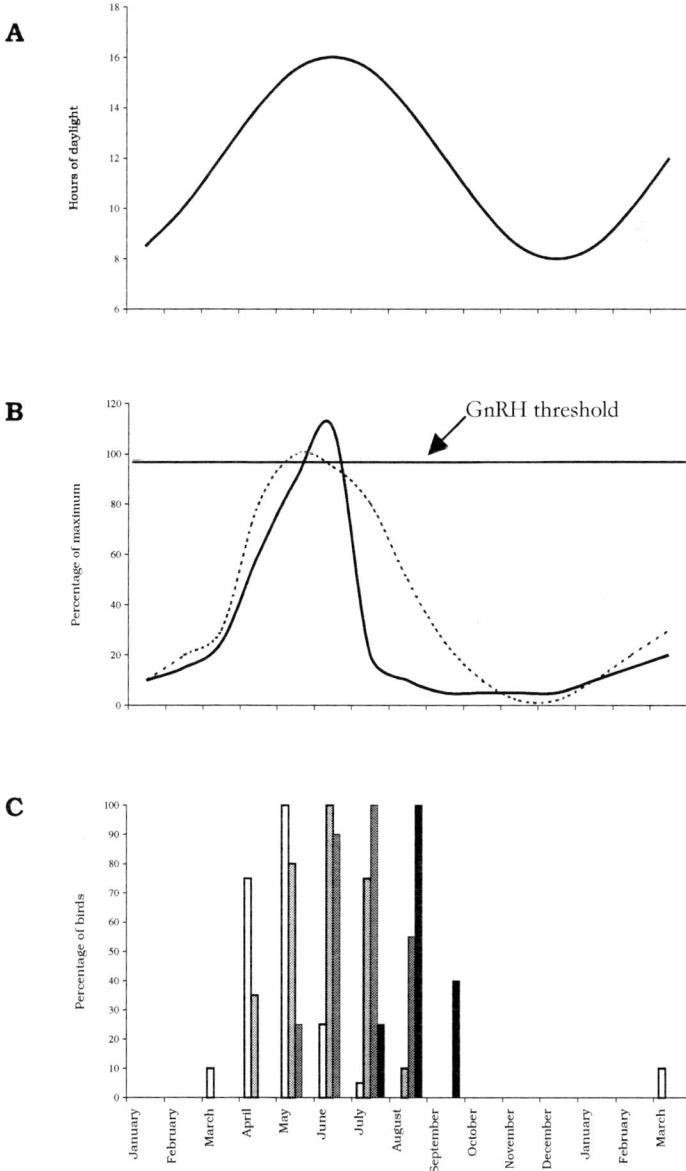

laying in many songbirds. The exact timing of breeding often reflects the availability of the food needed for egg production and feeding the offspring after hatching. For instance, the exact timing of breeding in European birds of prey depends on the availability of prey items. Kestrels breed early to take advantage of the ease of catching small mammals in the short grass of spring. By contrast, sparrowhawks feed on birds and breed slightly later in order to coincide with the fledging period of its main prey species.

Obviously many bird species live in parts of the world without marked changes in day length. In the tropics, hours of daylight do not vary over the year yet an annual pattern of gonadal development and regression is maintained although the changes are much smaller than seen in birds breeding at high latitudes. In general, the birds in the tropics remain ready to breed for much of year in order to take advantage of local factors, like rainfall, that favour breeding.

Mate selection and courtship

Breeding is often associated with birds establishing territories from which they can attract a mate. The role of the territory can vary between species and is basically defined as an area defended by the birds. The reason why the bird spends time and effort doing this can vary. In some species nest sites are at a premium and the bird defends an area where nesting is ideal. For instance, this may simply be the nest hole itself and not the surrounding area of forest. Other species defend feeding areas crucial for feeding chicks and defence of nesting sites is of little concern. In many species, however, the territory is important in providing a nest site within an area where food is plentiful.

Attracting a mate in birds invariably involves a visual and/or vocal display to advertise their willingness to mate and also their fitness as a parent. Birds, usually but not invariably the males, have developed a considerable range of feather structures and body adornments to highlight their sexual prowess. Many birds carry out prominent displays to impress the opposite gender. Dull-coloured songbirds often get around the problem of attracting a mate by having elaborate songs. Courtship in some species can also involve the male providing food for the female, a behaviour often seen in kingfishers and some songbirds.

Males in a variety of species, including many species of grouse and birds of paradise, have developed sexual displays at "leks" – arenas (this can be an area of land, a woodland clearing or the top of a tree) where local population of males come together in order to display to females. The males are out to impress, and subsequently mate with, as many females as possible. The reasons why females choose their mates is not always clear but once mated each female leaves the lek to build its nest, incubate and rear its clutch without the assistance of a mate. The males carry on displaying at the lek and if lucky father many clutches of eggs.

Breeding displays at leks are unusual in birds because the vast majority of species (over 90% of those studied) form a pair-bond between a male and female. Like in lek species, research has shown that it is the female who ultimately makes the decision regarding the pair-bond and copulation. This association may be rather short-lived in those species where the female rears the offspring alone but often it is prolonged by a need to share incubation duties or have two parents to forage for food for the hatchlings. In many species this pair-bond is reinforced throughout the period of nesting, incubation and rearing by elaborate displays when the birds meet. This is often important in colony-nesting species.

Nesting, incubation and rearing

The process of building a nest, copulation, laying eggs and then incubating them (Figure 1.2C) is largely covered by later chapters of this book. However, the egg and incubation by contact with the adult are key features of bird reproduction and their importance cannot be over-estimated. Should all go well and the birds have selected a good nest site, laid good quality fertile eggs and safely incubated them then they will produce a clutch of offspring that they have to rear to fledging.

Rearing birds to fledging, when they are independent typically being able to fly, often requires considerable effort from the parents. Their role largely depends on the developmental maturity of the hatchlings (see also pp. 49–51). Not all hatchlings are at the same stage of development and a spectrum in hatchling maturity has been recognised.

The megapodes of Australasia are at one extreme of this spectrum. Having dispensed with contact incubation, the parents bury their eggs in volcanic sand or mounds of rotting vegetation (see Box 5.2, p. 78) and their eggs and hatchlings are adapted to this lifestyle. The energy rich egg (see p. 38) allows a long incubation period and development of fully-feathered, "super precocial" chicks (Figure 1.3A). The young have to dig their way out of the sand or mound and are fully independent from the time they hatch. The parents play no role in the rearing of their offspring once they hatch. However, the hatchlings are so advanced that they can fly within a day of hatching.

Even though other birds, *e.g.* ratites, waterfowl (Figure 1.3B) and grouse, produce precocial young there is invariably a high degree of parental supervision in the days post-hatching. Fully precocial young are fully mobile and can feed themselves often learning from the example of their parents.

By contrast, there are many species, *e.g.* gulls (Figure 1.3C), that have semi-precocial young. These are fully feathered but are not very mobile and rely on

Figure 1.3. Appearance of hatchlings of A) a super precocial megapode, B) a precocial duckling, B) a semi-precocial gull, C) a semi-altricial owlet, and D) an altricial robin.

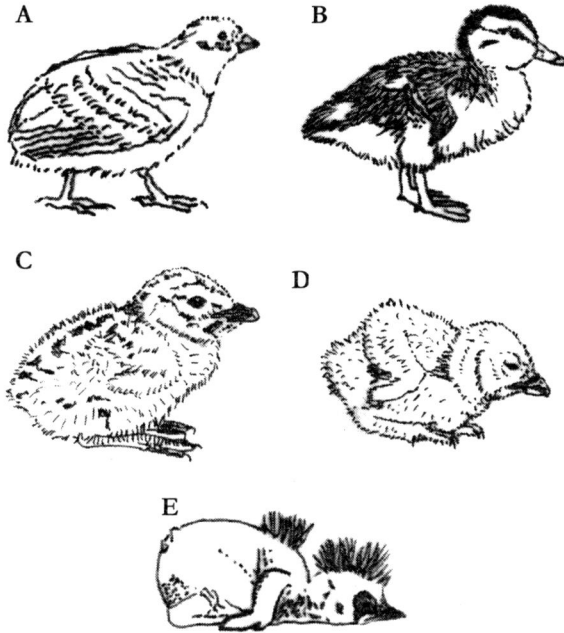

their parents to provide them with food. This is not surprising given that in many species the chicks would find it impossible to find food items like fish and buried worms.

Semi-altricial young of, for instance owls (Figure 1.3D) and penguins, are even less developed at hatching. They are covered with the lightest of down and usually need to be brooded by an adult to keep warm. Again the parents provide food.

Fully altricial offspring, typified by songbirds (Figure 1.3E) but also of pelicans, cormorants and gannets, are typically born naked and blind and totally helpless. The adults face a full time job providing sufficient food and warmth for their offspring.

The post-hatching period can be relatively short in birds because growth to adult size is typically quite rapid. In general, the period from hatching to fledging is similar to the incubation period. Once the offspring are fully feathered and fully independent then the reproductive season of the adults is complete and once there is a drop in the levels of the reproductive hormones and prolactin circulating in their blood many species enter a period of moult (Figure 1.1C).

Summary

- This book aims to describe both natural incubation nests as well as artificial incubation

- Breeding and incubation is under hormonal control ultimately affected by factors like day length but modified by factors like the prevailing weather

- Courtship often involves establishing a territory and a stable pair bond

- Offspring range from being highly developed, fully feathered and independent to being naked, blind and totally dependent on the parents

2 - Structural and Functional Aspects of Bird Nests

The nest plays crucial roles in reproduction in birds. Firstly, it acts as the receptacle for the eggs during incubation. The nest can protect the eggs not only by preventing them rolling away, but also by concealing them from potential predators. Secondly, the combination of the nest structure, and the bird sitting within it, forms the "incubation unit" essential for the creation of the optimal environment for development of the eggs. Post-hatching the nest serves to contain and protect the chicks from harm.

From the start it has to be realised that the vast majority of avian nests are quite unremarkable structures being simple cups or scrapes. Furthermore, nest morphology is usually species-specific and so I have refrained from describing in any great detail the range of weird and wonderful nests that birds choose to construct. Rather, the emphasis here is on the basics of nest construction and how these influence the functional aspects of the nest during reproduction. In this chapter, therefore, I describe the basic structure and location of nests in relation to these functions although the emphasis will be on the incubation phase. How nests are built and of what materials also feature. Finally, I describe how nests perform their roles in reproduction.

Where do birds put their nests?

Birds prove to be very adaptable in the positioning of their nests. It seems that for many small birds any crevice will do for a nest site and unusual nest sites, *e.g.* under car bonnets, in post boxes, *etc.*, regularly get reported. However, these remain the exception and most birds will adopt a natural nest site whenever possible. With two notable exceptions, nests are stationary structures constructed in one spot.

Recently Mike Hansell has analysed the nest building behaviour of birds and categorised the typical nest sites used. These, in no significant order, are: 1) in a tree or bush located at any height above the ground. 2) Built in and attached to grass or reeds at some point above the ground. 3) In a tree hole or a cavity excavated by the adult; large cacti can be included in this instance. 4) On the ground, and 5) in a ground hole or cavity. 6) Attached to the side of a cliff or wall. 7) On a flat ledge on a cliff face or wall, and finally 8) floating on water. Detailed analysis of nesting behaviour can make further distinctions within these categories but they are of no real relevance here. However, the choices of nest site and geographical location are going to have profound effects on the incubation environment and the relationships between the bird, its eggs and the nest type. For instance, grebes lay eggs on nests constructed on rafts of floating, usually rotting, vegetation and incubation often takes place with the eggs semi-submerged. This situation has meant that the eggshell has evolved to counter the wet conditions and yet lose sufficient water during incubation.

The exceptions I mentioned above are the Emperor and King penguins that do not have a fixed nest site. Rather the egg is incubated off the ground on top of the adult's feet under a "brood pouch" formed by a flap of skin. This means that the bird itself is its own nest site and is able to move around a little during incubation. Hence, these penguins have an incubation site rather than a nest site.

Factors affecting location of nest site

Although not widely studied a variety of factors have been recognised in the location of a nest site. For instance, it has been suggested that the European distribution of the song thrush is due to the availability of nest material. Absence of mud or dung and rotten wood prevents construction of the typical nest. No such restrictions appear to apply to the closely related and more widespread European blackbird.

Not surprisingly the prevailing climate is an important consideration for the bird making a nest. Exposed sites mean that the incubating birds, the eggs and perhaps the chicks have to suffer the prevailing weather conditions. In some species the sophistication of nest site location can be staggering. The Palestine sunbird nesting in Israel chooses its nest sites very carefully. Located within 2 metres of a sheltering structure 80% of them are located on the easterly aspect away from the prevailing wind and 70% of the nests only receive less than two hours of full sunlight. These sites minimise the effects of the climate on the nest environment – experiments have shown that wind can significantly increase cooling of the nests and eggs. Other species of bird will choose various aspects of a tree or bush in order to take advantage of, or avoid, prevailing weather conditions. Hummingbirds are at the mercy of the weather and choose sheltered nest sites that give protection from the cold night sky.

Many birds choose nest sites that provide some protection from predation. For instance, the colony nesting grounds of many seabirds not only reflect a scarcity of suitable sites but also the advantages of being part of a large group.

Many weavers and sparrows nest in colonies often building huge nesting colonies in trees. Location of a nest on a floating raft or high on a cliff face also reflects a degree of protection against terrestrial predators. Hornbills have gone to the extreme of excavating a nest cavity in a tree in which the female is imprisoned by all but blocking the cavity opening with mud. The male has to feed the female for the duration of incubation and chick rearing.

Many species adopt nest sites in close association with arthropods. Many kingfishers and parrots excavate nest burrows in termite mounds presumably taking advantage of the protection provided by the mound as well as the stable environment created by the termites. Many other species nest in close proximity to ants, bees and wasps. Several species appear to take advantage of the vigorous defence by wasps of their own nests. Although the presence of aggressive insects may deter predators of the avian nests there remains the danger that the birds and chicks will end up as victims of their potential protectors.

One critical factor in locating a nest is the presence of a food supply for the adults and the chicks. Most species are not very keen about travelling too far to get food as this can impose restricts on incubation or chick rearing. Indeed much territorial behaviour serves to protect the nest site and the local area for feeding (see pp. 6–7). Seabirds feeding way offshore are forced to nest on land but their incubation is prolonged because of extended times out to sea feeding. Nests are usually located in burrows and the eggs have adapted to long periods of chilling whilst the adults are off the nest and away at sea feeding (see Box 4.5, p. 64).

Having chosen a nest site the parent bird or birds have to build the nest before incubation can proceed. The location of the nest site and the materials chosen for nest construction strongly influence the shape and morphology of the nest.

Nest shape

Nests come in a variety of shapes although not all shapes can be found in all nest locations. The basic types of nest shape are shown in Figure 2.1. Cup nests (Figure 2.1a) are the most familiar type being constructed by a wide range of species and are characterised by a distinct bowl to hold the eggs. Dome nests have a roof over the cup forming a fully enclosed nesting environment (Figure 2.1b). Usually, entry into the dome nest is via a side entrance. An alternative structure is the dome and tube nest (Figure 2.1c) where entry into the nest is via a tube or an additional antechamber. A plate nest is often a large structure located above ground but with an indistinct bowl for the eggs (Figure 2.1d). A similar construction on the ground is a bed nest-shape (Figure 2.1e). Ground nests that lack any gathered material are called scrapes (Figure

Figure 2.1. Categories of nest shapes as defined by Hansell: a) cup; b) dome; c) dome and tube; d) plate; e) bed; f) scrape; g) mound; and h) burrow. See text for more details. Reproduced in a modified form from *Avian Incubation: Behaviour, Environment and Evolution*, with permission from Oxford University Press.

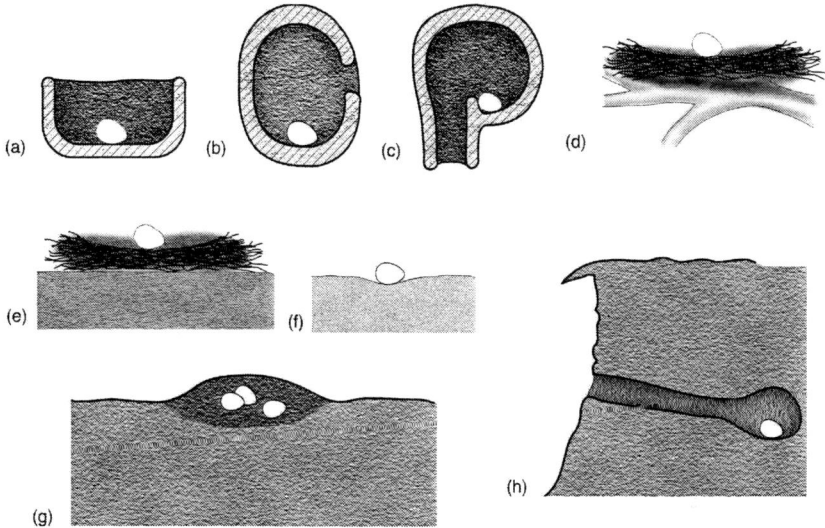

2.1f) and are typically a shallow depression on the ground. The megapodes have gone to extreme of building mounds for incubation (Box 5.2, p. 79; Figure 2.1g). A burrow (Figure 2.1h) is actually excavated by the bird in the ground, tree or cactus. By contrast, a hole nest is considered to be simply adopted rather than excavated by the birds.

These definitions are somewhat arbitrary but they do reflect the range of nest types. The complexity of the nest reflects on 1) the materials used to construct it and 2) the role played by the nest in maintaining the incubation environment.

Interestingly there is clear relationship between the size of the bird and the weight of its nest. Small birds produce light nests relative to their body weight whereas big birds tend to build large nest constructions that far exceed their own weight. In birds of prey nests are often re-used every year and over time the nests simply get bigger until the supporting structure can hold them no longer.

Composition and construction of nests
Before discussing the techniques employed by birds during nest construc-

tion I will briefly describe the materials birds employ in the various structural layers of their nests. Thereafter I will briefly describe the techniques that birds employ to build a nest that will survive the incubation (and rearing) period.

Nest materials

A wide variety of materials have been found within bird nests and can be largely characterised into three main types (Table 2.1).

Animal materials are mainly feathers and fur although snakeskin is used by some species to cover the nest. Swallows, martins and swifts use salivary mucus in order to build nests by sticking other materials together. Some swiftlets rely entirely on salivary mucus to build the nest. The other widespread animal material is the silk produced by caterpillars and spiders. This is used to hold other nest items together although some hummingbirds use silk strands to secure the nest to its attachment.

Not surprisingly birds employ a wide variety of plant materials (Table 2.1). These can be found in most of the layers of the nest and may play other important roles in maintaining nest hygiene (Box 2.1, p. 15). Different plant types lend themselves more to other roles in nest construction. For example, grasses are easier to weave with and sticks will form bulky nests more readily.

Table 2.1. Types of material employed by birds in nest construction.

Type of nesting material	*Examples*
Animal	Feathers, mammalian fur and hair, snake skin
	Salivary mucus
	Silk from insects and spiders
Plant	Leaves and associated leaf structures, including pine needles
	Woody stems and sticks
	Rootlets and vine tendrils
	Plant down
	Grasses
	Palms
	Flower heads
	Horsehair fungus
	Ferns, mosses, lichens
	Rushes
	Seaweed
Mineral	Mud
	Stones
Other	Cow dung
	Paper, plastics and other synthetic materials

Of the mineral materials used for nest building mud is very popular. 5% of bird species use mud in the various layers of their nests. Stones are used by Gentoo penguins to raise their nest site above the ground thereby preventing water logging of their eggs. Other materials found in nests appear to be usually of human origin and birds appear to take advantage of structural properties similar to more typical materials.

Structural zones of nests

Four main layers of a nest have been recognised based on their function. Not all nests have all of these function parts. In particular, the scrape nests of the ostrich are depressions in the soil and many shorebirds lay their eggs surrounded by pebbles on the beach with no additional material. Other birds, such as guillemots, simply nest on bare ledges without any nesting material.

Nesting materials are built into one (or more) of the four structural layers, each with a different function. "Attachment" describes the layer that is in

---

BOX 2.1 – MICROBIOLOGY OF NESTS

Nests create the perfect micro-environment for embryonic development. Unfortunately, the combination of warmth and elevated humidity is also ideal for the growth of microbes. Research has shown that there are thriving populations of both fungi and bacteria within nests. Many of the fungi are derived from, and live on, the feathers of the adult birds. The range of bacterial species is wide and again is usually derived from the parents. Some of the species of microbe isolated are pathogenic but the extent to which they cause embryonic mortality or parental disease is not known.

Thought to be an attempt to counter microbial growth in the nest (and perhaps deter arthropod parasites) many species incorporate green plant material into the nest lining or construction. Although this may provide camouflage or shading for the nest, experiments have shown that at least some of the plant species involved have anti-microbial properties. There is also some suggestion that waxy secretions used by the bird for preening have an anti-microbial role.

The presence of bacteria in a nest may be unavoidable but in some waterfowl it may well be a real boon. Research has shown that the bacterium *Bacillus licheniformis* is able to grow and thrive on the waxy cuticle of the eggshell even excluding other bacterial species present on the shell. Indeed, in the Mandarin duck the bacteria gradually degrade the cuticle leading to an increase in water vapour conductance. How widespread this is in waterfowl, and birds in general, is not known.

Figure 2.2. Examples of the type of nest attachments as described by Hansell. a) top; b) three variants of top lip; c) top side; d) bottom side; e wall; f) bottom multiple (branched); g) bottom multiple (vertical); h) leaf purse; i) bottom; and j) on ground. Reproduced in a modified form from *Avian Incubation: Behaviour, Environment and Evolution*, with permission from Oxford University Press.

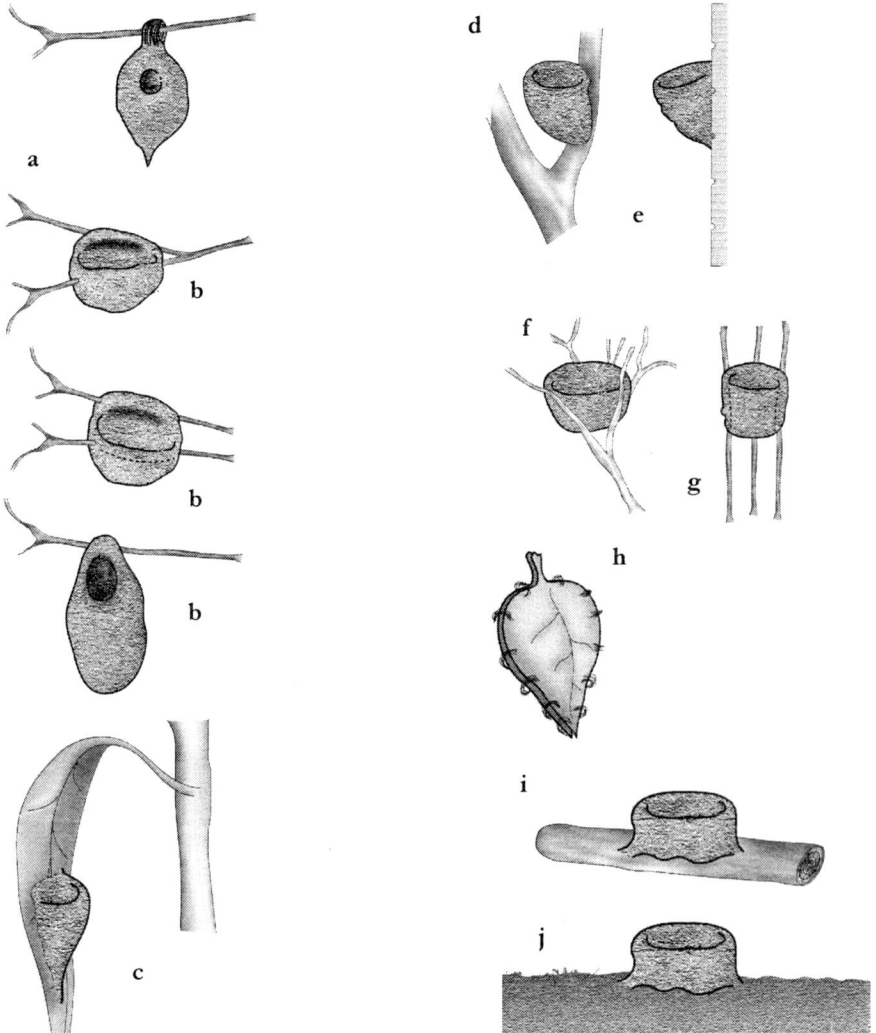

Figure 2.3. A long-tailed tit nest. Note the way the structural layer is woven around the tree branches. The entrance hole is at the top.

volved with securing the nest to the substrate. Those nests that are not supported from below require a mechanism to keep them in position. The variety of these attachment types is shown in Figure 2.2. Nests can be attached by weaving materials around supporting structures (Figure 2.2b) or they can be attached to supporting structures by adhesive materials such as saliva, mud or arthropod silk (Figure 2.2.i).

The "Outer" layer forms the external surface of the nest and is deemed to be a decorative layer that alters the appearance of the nest without affecting its strength or acting as an attachment. This layer of the nest is typical of nests of small birds and is absent in those of bigger birds. Many of the materials employed on the outside of nests are silk and lichens, which may serve to break up the outline of the nest. Alternatively, the lichens on the long-tailed tit nest (Figure 2.3) may act as a waterproofing layer.

The "Structural" layer forms the bulk of many nests. Providing integrity to the nest structure, the role of this layer is to support the eggs and bird during incubation. In larger nests, the structural layer forms the whole of nest and has to adopt some of the functions of the attachment and lining layers. In general, as the bird size increases then the material used in the structural layer increase in strength. Therefore, very small birds can use spider silk to build nests, larger birds use grasses but crows and other bigger birds need to employ sticks in the structural layer.

The "Lining" layer has no structural role and lies adjacent to the eggs although many nests do not have lining layers. A variety of plant and animal materials can be used in the lining layer but only feathers and fur are employed exclusively there where they prove very useful for their high insulation properties.

Construction techniques

Birds employ a variety of techniques for nest construction ranging from the very simple to the extremely complex. In essence there are two ways of building a nest: 1) removal of material or 2) assembly of material.

Removal of material can simply involve scraping soil or sand away from a part of the ground to form a nest scrape. Other types of removal of material involve sculpting of a cavity by removing substrate from the ground or tree (Figure 2.1h). The obvious example is the woodpeckers, which have extended their normal feeding behaviour of chiselling away at bark and wood in order to excavate a nest hole. Other birds have developed sculpting as a secondary skill. For example, kingfishers and bee-eaters dig nest burrows in banks of soil, a behaviour far removed from their normal feeding behaviours.

Moulding involves birds using mud to stick together other materials to construct a nest structure. The parents mould the final shape of the nest as they proceed. Swallows and martins build up a nest with thousands of mud pellets usually including some grasses or hair to add some strength. It is curious that these nests are built on rock faces or walls without any support from below (Figure 2.2e). Other species employ saliva as well as mud as the adhesive for other materials such as plant material and feathers. By contrast, the edible-nest swiftlet of bird's nest soup fame, uses saliva alone to build the nest. Alternatively, mud or salivary mucus is often used to plaster the inside of a nest cup. These adhesives are used to give structural integrity to nests of woven plant material in many species of bird.

Piling up of nest material can raise the nest above the substrate and on the ground this can involve plant material (*e.g.* the grey heron) or stones (*e.g.* the Gentoo penguin). For many ground nesting birds it is usually sufficient for the nest material to be piled up because the ground gives a lot of support to the structure; nest cups are made by moulding the nest material with the feet and body. In a tree piling up involves construction of a platform of sticks and twigs that eventually form a nest cup and is seen in a wide variety of species including herons, storks, birds of prey, crows and pigeons. The structure has to be substantial enough to ensure that it does not fall from the branches. In many species this involves a simple process of laying sticks on top of each other with no attempt to interlace them. However, many species are adept at manipulating the twigs to intertwine them into a more ordered framework.

Interlocking construction behaviour relies on the skill of the bird to build a nest from dry materials and without the aid of adhesives such as mud or saliva. In many examples this behaviour is simply an extension of the manipulation of twigs to strengthen a structure. Many species intertwine grasses and leaves around supporting structures as attachments and then interweave the grasses to form the nest cup (Figure 2.2g).

A few bird species have developed a technique of sewing leaves together to form a pouch (Figure 2.2h). Arthropod silk and plant down are used as the thread. Spiderhunters have developed a technique of pop-riveting leaves together using spider silk thickened at one end and then forced through the leaf!

Many small birds employ a system of holding together nest materials by hooking arthropod silk around plant materials (in a similar way to "Velcro"). Long-tailed tits build domed nests (Figure 2.3) containing more than 600 spider egg cocoons teased out to form a fluffy mass that is hooked onto other nest material. This system is very effective because it is simple to achieve a complex nest form and individual parts of the nest can be readily taken to pieces and re-fixed as necessary.

Weaving involves the complex interlacing and knotting of grasses and leaf strips to form complex nest structures that typically hang from branches. The techniques employed by the Old World weavers and the New World oropendolas, caiques and orioles are remarkably similar.

Having described the building techniques employed by birds it is interesting to pose the question of how easy is it for a bird to construct a nest. For most birds nest location is a critical factor determining the type of nest that can be built. Thereafter, simple stereotypical behaviours can be employed to construct a nest and most birds will position themselves where the nest cup will be and then construct it around themselves. Large nest platforms composed of piled up sticks requires little complex behaviour and the use of materials like arthropod silk can mean that small birds can construct quite complex structures by pushing nest materials together. Whether nest construction techniques are innate or learnt from watching others and from experience has not yet been fully resolved.

Functional aspects of the nest during incubation

Natural incubation depends entirely upon the inter-relationship between the bird sitting on eggs in a nest. Nest type is highly species-specific and so its role in incubation cannot be underestimated. The type of nest constructed and the materials used, almost certainly reflect the particular requirements of the incubating birds and eggs within the chosen nesting environment. Hence, attachment layers of nests are important in securing the construction in place for the duration of the incubation and nestling periods. Outer layers may serve to camouflage the nest or provide water proofing. Structural layers ensure that the nest can support the bird, eggs and chicks. Choice of the lining layer affects the insulation of the nest cup and may be intimately involved in maintaining egg temperature or gaseous environment.

Very little work has been done on the thermal characteristics of nests but it has been shown that some species build nests with higher thermal insulation than others do. How this reflects the prevailing climate or affects nest attentiveness (see pp. 70–77) remains unclear. Furthermore, the gaseous environment can be significantly affected by the nest construction. The open stick nests of pigeons are likely to have higher values for gas conductance than the hole nests of parrots. This will also affect rates of ventilation of the nest. For instance, in the burrow nesting bee-eater the movement of the bird up and down the burrow leading to the nest chamber is essential in maintaining appropriate levels of oxygen and carbon dioxide (see also pp. 92–93).

Nest construction and location can affect predation rates of the eggs or chicks. How nests are built may not only reflect the need for incubation of the eggs but also be significantly involved in ensuring high survival of the chicks. The scrape nest of the ringed plover on a beach is sufficient for incubation of the eggs but plays no role in rearing the chicks, which wander away from the nest site soon after hatching. By contrast, the nest cup of reed warblers is constructed attached to reeds. It not only has to retain the eggs but also has to be deep enough to prevent the chicks from falling into the water below.

Our understanding of the roles nests play in maintaining the incubation environment is very poor yet it could provide important insights into artificial incubation. It is hoped that this aspect of bird reproduction will receive more attention in the future.

Summary

- Most nests are unremarkable cups or scrapes
- Nests can be located in a wide range of places influenced by availability of nest materials, prevailing climate, predation and food supply
- Nest shape varies according to location and nesting material
- Nesting materials are primarily plant derived but include animal and mineral materials
- There are four layers of a nest, each with different functions
- Nests can be constructed by removing material or by assembly of materials
- Nest construction techniques range from simple scraping away of soil to complex weaving of grasses
- The nest is critical in the formation of the incubation environment

3 - Egg Formation, Structure and Function

Although not restricted to birds, the rigid-shelled egg does help define bird reproduction. It is hard to envisage how contact incubation could have evolved if the eggshell had not been able to withstand the presence of the adult. This chapter describes eggs in some detail providing information on three key areas of oology (the study of eggs). The first is the formation of the egg. The second section provides information on the basic structure and size of the egg, including the eggshell. The final section deals with the chemical composition of eggs.

Egg formation

Egg formation begins with the maturation of an ovum within the ovary. The expansion of the ovum into the yolk takes places over several days and involves cyclical deposition of vitellogenin (a lipid-protein complex) manufactured within the liver and transported via the blood to the ovary. Dosing the hen with a fat-soluble dye every other day produces yolks that when cut in half exhibit rings of dyed and normal yolk. In the domestic fowl yolk deposition takes 9–10 days but information for other species is harder to obtain. In the fairy tern a 7–8 g yolk takes around 8 days to complete compared with 25 days for a 90 g yolk of an albatross. It would be interesting to get more information on the rate of yolk formation in smaller species.

Once the yolk has achieved the appropriate size and meiotic division (see Box 4.1, p. 44) is completed the ovarian tissue surrounding the yolk ruptures and releases it into the peritoneum. This process is under hormonal control and in the domestic fowl it takes place ~25 hours before the egg is laid. During the breeding period of the female there is a continuous sequential production of ova. Once one yolk is released then the next largest ovum enters the final stages of maturation before it is released 1–2 days later. Note that in birds only the left ovary and oviduct are functional.

Egg formation takes place in the oviduct, an open-ended tube extending from the cloaca up to the ovary located next to the kidneys (Figure 3.1). The oviduct does not connect directly with ovary but rather opens into the peritoneal cavity of the abdomen. Given that the other end of the oviduct opens into the cloaca, which is shared with the openings of the intestine and ureters,

Figure 3.1. Location of the left ovary and oviduct within the abdomen seen in cross section from the left (A) together with the principle components of the oviduct (B).

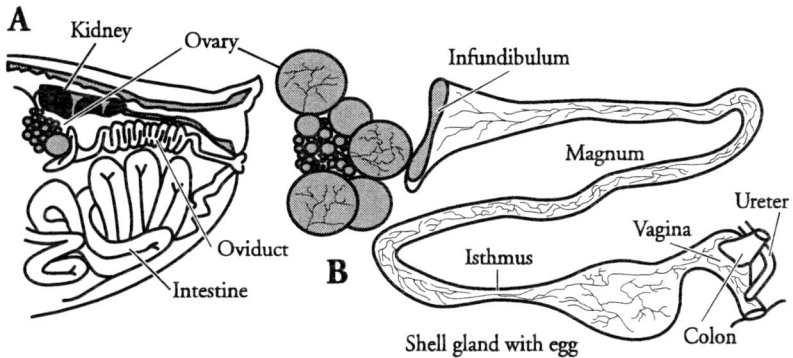

it is fascinating how the oviduct and the peritoneum remain free from bacterial contamination.

The yolk is actively engulfed by the infundibulum at the top of the oviduct and this is the site of fertilisation (see pp. 44–45). The fibrous inner peri-vitelline layer that surrounds the yolk is supplemented by fibres of the outer peri-vitelline layer secreted by the wall of the infundibulum. The chalazae are also formed after the vitelline membrane is completed and twisted as the yolk slowly rotates in the oviduct. The yolk spends less than 30 minutes in this part of the oviduct before moving into the magnum.

Albumen proteins are deposited around the yolk during the 3 hours the yolk spends in the magnum (Figure 3.1). The proteins are secreted by the wall of the magnum in a dehydrated form and form a capsule around the yolk.

The yolk-albumen complex then moves into the isthmus (Figure 3.1) where it stays for another 1½ hours. Here glands in the wall of the oviduct secrete the fibres of the shell membranes. The fibres are interwoven to form a mesh. By the end of its time in the isthmus the surface of the shell membranes is flaccid and has an area larger than that produced by the volume of yolk and albumen present at that time.

The yolk-albumen complex spends 20 hours in the shell gland (Figure 3.1). Here two processes take place. In the short first part of the shell gland, there is transfer of water and mineral salts ("plumping") into the albumen capsule. Plumping causes the albumen proteins to expand increasing the egg volume and making the shell membranes go taut. In this way the egg assumes its final size and shape. This takes approximately six hours in the domestic fowl egg.

The second process occurs in the shell gland pouch and is the growth of calcium carbonate ($CaCO_3$) as calcite crystals from "seeding sites" laid out at random on the outer surface of the outer shell membrane. Crystals grow (see also pp. 27–28) in a medium rich in calcium and bicarbonate ions that is secreted by the walls of the oviduct. Shell deposition takes most of the time during the egg's stay in the oviduct but shell strength only begins to increase after 8–10 hours in the shell gland. After shell deposition is completed the egg is rotated within the shell gland prior to laying.

Pigments are produced and secreted by the cells lining the shell gland and are deposited during the last few hours of shell formation (Figure 3.1). Once the shell is completed the shell accessory material (cuticle or cover) is deposited on the outer surface of the vertical crystal layer. Immediately prior to laying the egg is pushed into the vagina in preparation for oviposition through the cloaca and vent (Figure 3.1).

The timing for the periods spent in the various parts of the oviduct are for the domestic fowl egg. Very little research has been carried out on egg formation in other species. It can only be assumed that the pattern of formation is the same in most, if not all, species and that it is only the actual length of time in each section of the oviduct that varies. This assumption can be made because of the close similarity between the structure of eggs from a wide range of birds.

Size and structure of eggs

Bird eggs are essentially the same irrespective of their size. From the smallest hummingbird to that of the ostrich, the structure of the egg is the same. There are, of course, differences in the dimensions and shape, and composition of eggs (see pp. 37–41).

Mass, dimensions and shape

Egg mass can be described as a function of the body mass of the female. In general as body mass increases then egg mass gets bigger at a slower rate – a ten-fold increase in female weight is reflected by only a 5 fold increase in egg mass. As a result egg mass as a percentage of body mass decreases (Figure 3.2) with ostriches having the smallest eggs relative to body size (1.25–1.5%). However, there is considerable variation in the relative size of eggs between different orders of birds. For instance, a 100 g parrot will lay a 6.4 g egg compared with the 17.4 g egg laid by a shorebird of the same mass.

Eggs come in a range of sizes with the average ranging from 0.5 g in some hummingbirds up to 1,500 g in the ostrich. However, there is considerable variation within a species, for example the ostrich can lay viable eggs in the range from 1 kg to over 2 kg. The dimensions of eggs scale to egg mass and volume. These relationships can be used to determine fresh weights of eggs

Figure 3.2. Relationship between body mass and the relative size of bird eggs.

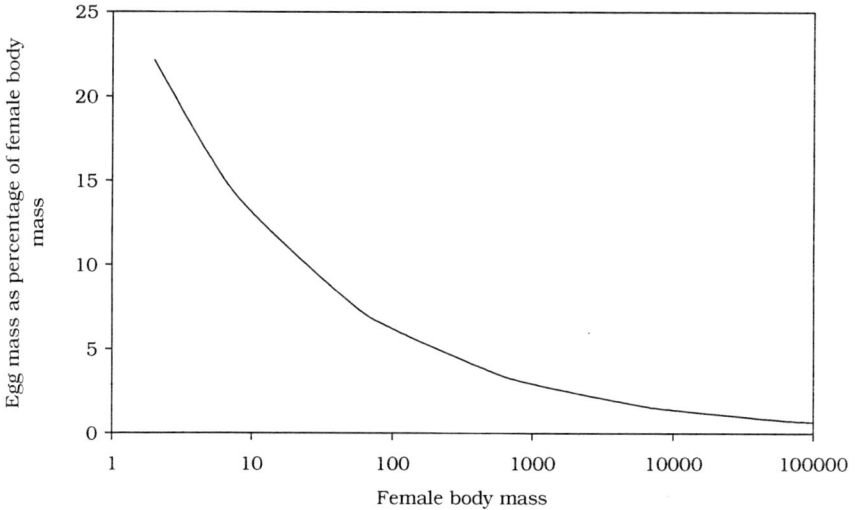

for which initial mass is unknown (Box 3.1, p. 25). The smallest eggs have a maximum breadth of ~8 mm compared with ~130 mm in the ostrich. The extinct *Aepyornis* elephant birds of Madagascar laid the largest eggs produced by a vertebrate. These measured around 260 mm in diameter and weighed well over 10 kg!

Egg shape can vary considerably and is often defined by the elongation ratio between maximum length and maximum breadth of the eggs. Round eggs typical of hole-nesting species have elongation ratios of only 1.2 (*i.e.* length is only 20% longer than breadth) but long pointed eggs have ratios of 1.5 or more. Typically bird eggs are asymmetrical with one end of the egg (typically the end at which the air space forms) being broader than the other. The big eggs of ratites, and rounder eggs, tend to be more symmetrical. The function of egg shape is not clear but in many species asymmetry helps to form compact clutches of eggs in the nest and maximises contact with the brood patch.

Egg shape is caused by differential muscular contraction in the oviduct with the blunt pole of the egg being formed immediately after the point at which the egg is being pushed down the oviduct. Hence, the egg proceeds down the oviduct sharp end first but prior to laying it is rotated through 180° within the shell gland and is laid blunt end first.

Structural aspects of albumen and yolk
The spherical yolk is located centrally within the albumen. Although the

---

BOX 3.1 – DETERMINING FRESH EGG MASS AND VOLUME FROM LINEAR DIMENSIONS

It is sometimes necessary to determine the fresh weight of an egg at some time after incubation has started or after a prolonged period of storage. This is easy to do by measuring (in cm) the maximum length (L) and maximum breadth (B) of the eggs with vernier callipers. These values can be used in the following equation to produce an estimate for fresh eggs mass (IEM, in g):

$$IEM = 0.548 \times L \times B^2.$$

The volume (V, in cm^3) of an egg can also be predicted from L and B:

$$V = 0.507 \times L \times B^2.$$

The constants, 0.548 and 0.507, are averages for a range of bird species and egg sizes. If dealing with a particular species it is often useful to determine the values from fresh eggs. For instance the weight constant for Houbara eggs is 0.552.

Alternatively egg volume, and shell surface area, can be calculated from egg mass alone:

$$V = 4.951 \, IEM^{0.666}$$

$$SA = 4.835 \times IEM^{0.662}$$

---

vitelline membrane is continuous with the chalazae it is unlikely that these actually support in the yolk in any meaningful way. Rather it is the surrounding albumen that holds the yolk, which tends to float towards the upper part of the egg. Changes in albumen structure before and after incubation allow the yolk to move up nearer to the upper part of the eggs.

As indicated earlier the yolk is a layered structure composed of "yellow" and "white" yolk (Figure 3.2). The latter is a smaller part of the yolk but it is more watery having a lower density than "yellow" yolk. Moreover it is located in the upper part of the yolk and immediately below the embryonic tissue the "white" yolk extends as a tube down into the core of the yolk (Figure 3.2). This arrangement means that the upper and lower hemispheres of yolk differ in their densities. If the egg is turned then the denser lower hemisphere always brings the top hemisphere, with its associated embryo, back to the top of the egg.

Albumen surrounds the yolk in a series of layers. Immediately next to the vitelline membrane there is a very "thin" layer of "thick" albumen surrounded by a thicker layer of thin albumen. The distinction between these layers is due to the amount of the protein ovomucin within the complex of albumen proteins – thick albumen is much richer in ovomucin. The bulk of the albumen within the egg is thick albumen and there is a layer of thin albumen ringing the

Figure 3.3. Diagrammatic representation of a bird egg in longitudinal section through the midline of the egg.

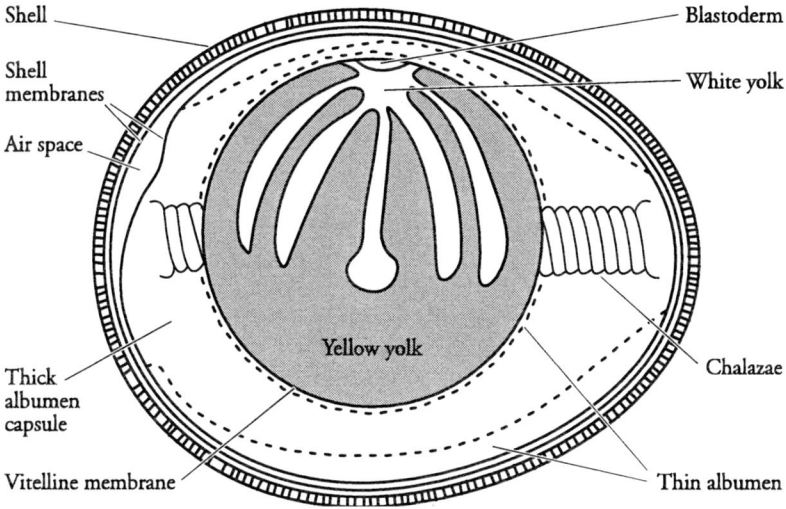

inside of the eggshell (Figure 3.2).

Embryonic development changes this arrangement during incubation (see pp. 55–56) but before incubation prolonged storage also has dramatic effects on albumen structure and composition. During egg formation the egg is in an environment rich is dissolved carbon dioxide (CO_2) and so the egg contents are rich in dissolved CO_2. Once in the nest (or egg store or incubator) the high concentration of CO_2 in the egg starts to dissipate by diffusion of this gas through the eggshell pores. CO_2 is important in controlling the acidity of bodily tissues and its loss from the egg lowers the acidity (pH) of the albumen. Furthermore, as the time between laying and incubation extends then the ovomucin begins to breakdown and the thick albumen becomes thinner. The end result is clearly seen when fresh and stored fowl eggs are fried. A fresh egg has the albumen forming a mound around the yolk and there is little run-off of thin albumen. By contrast, a stored egg has much more thin albumen and the albumen spreads out to form a thin layer of cooked white. Within the egg this thinning process removes the restraint around the yolk and it is free to move nearer to the top of the egg.

Recently there has been work looking into the relationships between egg size and composition. Basically, is it the yolk or the albumen that determines the length and breadth of the egg? The evidence to date suggests that both components are important in determining egg breadth but the amount of albumen in the egg is more critical when determining length.

Eggshell function, formation and structure
The bird eggshell is one of nature's greatest wonders. It fulfils a wide range of roles including: 1) camouflaging the egg in the nest, 2) physical protection of the internal egg contents against injury and microbial attack, 3) regulating gaseous exchange between the egg contents and the ambient environment, 4) conduction of heat from (and to) the brood patch (and air), and 5) storing calcium ions for use by the embryo during development. All this is made possible by a relatively thick layer of calcium carbonate crystals.

The shell structure reflects the pattern in which it is formed. Starting from the innermost layer adjacent to the albumen there is a very thin "limiting membrane" a continuous layer of protein underlying the inner shell membrane that is in turn covered by the outer shell membrane (Figure 3.3). These two membranes are made of fibres of protein wrapped around the yolk and albumen during the egg's time in the isthmus (Figure 3.1). The two membranes differ in the thickness of the fibres with the outer shell membrane forming a thicker layer of coarser fibres. An air space is created between these two membranes as the egg loses water vapour and the contents shrink. The hard shell is deposited on the outer surface of the outer shell membrane.

The hard part of the shell is made of crystals of $CaCO_3$ deposited as calcite ($CaCO_3$ can form various crystal forms and calcite is commonest) in the shell gland. During the first stages of crystal formation nodules of calcite form on the outer shell membrane and crystals grow outwards from these "seeding sit-

Figure 3.4. Diagrammatic representation of the basic structure of the avian eggshell showing the variety of pore types observed (A–D). Arrows at E indicate the pattern of calcite crystal growth (see text). OSM = outer shell membrane and ISM = inner shell membrane. Note that the cuticle is not shown in this illustration.

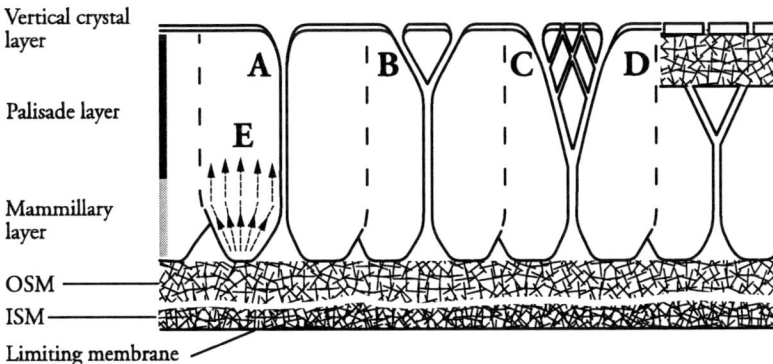

es" to form a mammillary cone (see arrows in Figure 3.4). Eventually the crystals from adjacent seeding sites meet each other and the only way for the crystals to grow is on the outermost surface. The crystals forming the palisade (or spongy) layer grow out from the mammillary layer and form the bulk of the shell (Figures 3.4 and 3.5). There is typically an organic matrix that permeates the calcitic shell providing a framework for crystal growth. In many species a vertical crystal layer delimits the outer surface of the eggshell where the orientation and size of the crystals change. This marks the point at which the correct shell thickness has been achieved and crystal growth needs to stop. The calcitic shell is also permeated by pores (Figure 3.4), which allow for gas exchange between the egg contents and the nest environment (see pp. 31–32).

Shell thickness scales with egg mass varying from only a few micrometres in small eggs (*e.g.* 50 μm in the European blackbird) to 2 mm in the ostrich eggshell and 3.5 mm in *Aepyornis* eggshells. Research has also shown that eggshell strength scales with body mass of the adult birds such that the shell will support the weight of the bird sitting on it. However, the degree of leeway in this relationship is much smaller in big eggs. Hence, the problem of breakage of eggs due to pesticide-related thinning of the eggshell has been to be more prevalent in the bigger eggs of herons and birds of prey.

The outermost layer of the eggshell is the shell accessory material, often called the "cuticle" although this more properly refers to a layer of organic material. "Cover" is the term used for a layer of inorganic salt crystals. Shell accessory material can come in the form of small spherules or as an amorphous mass of material (Figure 3.6).

Shell accessory material acts as the first defence against water and potential water-borne contaminants of the egg contents. The extra layer blocks the openings to the pores preventing fluids from entering the pore canal. The material can simply lie over the pore orifice or can extend down into the pore opening forming a plug (Table 3.1).

Either way the seal is not complete and there is free movement of gases in and out of the pore. The eggs of many species of birds, *e.g.* ostrich, emu and pigeons, do not have any shell accessory material on the external surface of the shell. The nesting conditions of these species are often very dry and the danger of water coming in contact with the shell is remote. By contrast, the waxy cuticle of waterfowl eggs repels water very effectively helping to protect the egg contents from bacterial contamination.

In some species the shell accessory material can form an additional barrier to gas movement and reduce the gas conductance of the shell (see pp. 33–34). Removal of shell accessory material has been shown to increase the porosity of the eggshell and water loss is higher than it would have been. In some penguins and the Houbara bustard the shell accessory material is rubbed away du-

Figure 3.5. Scanning electron micrograph of the eggshell of the Houbara bustard showing the extent of the different crystal components of the shell. The arrow indicates the point at which a mammillary knob is attached to the outer shell membrane. Photograph by D. C. Deeming & G. K. Baggott.

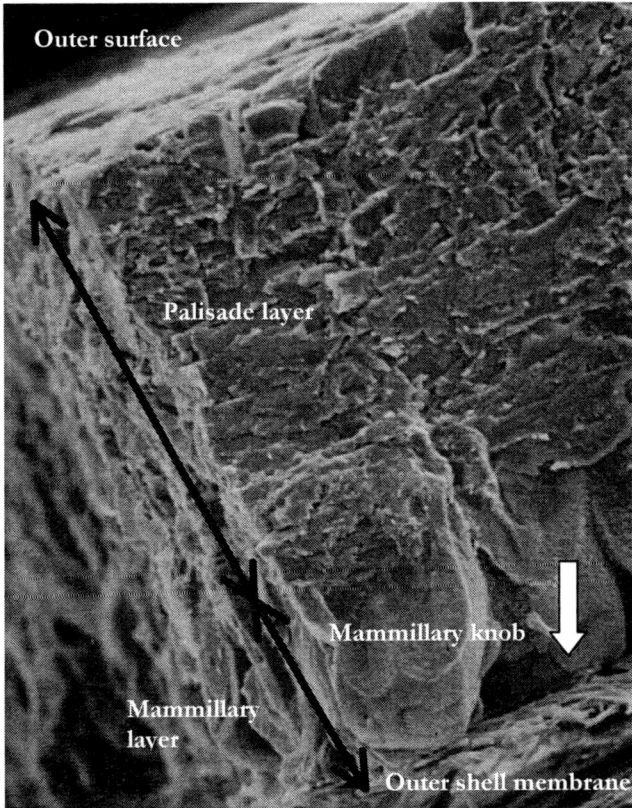

ring natural incubation and weight loss increases as incubation progresses. In the mandarin duck, water loss is increased by degradation of the cuticle by a species of *Bacillus* bacteria (Box 2.1, p. 15). Whether erosion of the shell accessory material is widespread in birds has yet to be fully investigated.

Egg colour is a key aspect of the role of the eggshell. Pigment laid down in the calcitic shell and/or in the cuticle gives the eggshell a pattern of background colour and spotting often distinctive to species. The exact pattern of spotting or background colour can vary between individuals and within a clutch. The role of egg colour is not well understood but it has roles in egg recognition (Box 3.2, p. 31). In ground-nesting species, colour and spotting act to camouflage the egg in the nest. By contrast, hole-nesting species usually

Figure 3.6. Scanning electron micrographs showing the appearance of the cuticle of A) red-legged partridge and B) Houbara eggshells. Compare the "cracked mud" appearance of the spherulitic cuticle in A with the flat featureless cuticle in B. Scale bars are 100 μm. Photographs by (A) Alex Fraser & Maggie Cusack and (B) D. C. Deeming & G. K. Baggott.

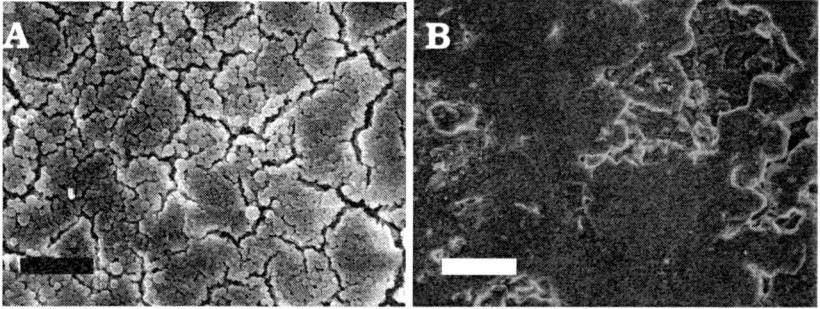

Table 3.1. Variety of pore types and the relationships with shell accessory material.

Shell accessory material (SAM) coverage	Pore canal	SAM type	Examples
None – *Pores open into honeycomb-like structure under the external surface*	Unbranched		Storks
	Branched and unbranched		Emu and cassowaries
None – *outer pore orifice open*	Unbranched		Pigeons
	Branched and unbranched		Ostrich
SAM forms a skin on shell surface with cracks over pore openings – *pore orifice occluded*	Unbranched		Gulls
	Branched and unbranched		
SAM forms a plug within the outer orifice – *outer pore orifice plugged*	Unbranched		Lily-trotter
	Branched and unbranched		Rhea
SAM covers the shell surface capping and sometimes blocking the pore openings – *pore orifice capped*	Unbranched	Organic Inorganic	Swans Cormorants
	Branched and unbranched	Organic Inorganic	Penguins

have white eggs, which are presumably easier to see in the gloom of the nest hole. In at least one instance egg colour has been shown to reflect the ultrastructural morphology of the eggshell (Box 3.3, p. 32).

Pores

Although the eggshell appears to be a solid structure it is riddled with pores – air-filled tubes that lie between blocks of calcite crystals. These are critical for the survival of the embryo as they allow diffusion of oxygen into the eggs and carbon dioxide out the egg. Water vapour also leaves the egg through the pores.

It is not clear how pores are created in the shells but it is likely that they form at random at places where the growing calcite crystals of adjacent mammillary knobs do not fully abut each other. The space created is maintained as the palisade layer grows outwards. This pattern of crystal growth appears to favour development of a "post horn" shape for the pore – simple tubes topped by a funnel shape (Figure 3.4A). This is the pore morphology observed in the majority of species but even within a single eggshell this basic pattern varies in the diameter of the pore and the depth of the funnel at the pore orifice (Figure 3.7). Thicker eggshells have branched pores (*e.g.* in the rhea eggshell; Figure 3.4B) with the branching being more complex as shell thickness increases (*e.g.* the ostrich eggshell; Figure 3.4C). The eggshells of the emu and

BOX 3.2 – EGG RECOGNITION

To date the available evidence suggests that egg coloration does not serve to allow the parent birds to recognise their own eggs. Although female ostriches are reported to recognise their own eggs in a nest of eggs laid by several females, it is unclear how she distinguishes the cream-white eggs apart. In most other species egg recognition is rare. Many host species of brood parasites (see Box 5.1, p. 68) are unable to recognise parasite eggs despite differences in size and colour. Many species of bird without a defined nest structure are able to retrieve eggs that roll from the nest scrape but they will bring in any egg irrespective of whether it is theirs or not. Many other species will ignore their own eggs even though they are a few centimetres away.

Experiments have shown that the urge to incubate is very strong in many species. Individual birds will accept and sit on a variety of objects irrespective of whether they resemble eggs or not. In some instances the larger the "egg" then the greater the stimulus to sit on it even if the object was too big for the bird to sit upon.

Marking eggs to study egg turning behaviour (see pp. 94–98) adversely affects incubation behaviour in some species. However, the willingness of birds to sit on a variety of objects has been invaluable in the use of telemetric and other artificial eggs to study incubation (see Box 5.7, p. 97)

BOX 3.3 – SHELL COLOUR AND HATCHABILITY.

In commercial pheasants eggshell colour can be used to predict whether eggs will hatch. It has been long recognised that blue eggs do not hatch as well as eggs with the more typical brown or olive coloured shells. Recent analysis has revealed why this is the case.

A study of the ultrastructure of the eggshell in blue and brown eggs showed that blue eggshells are thinner and the connection to the outer shell membrane is defective. The blue shell is of poor quality and appears to be laid prematurely. High embryonic mortality associated with these shells is due to high weight loss during incubation. The thinner than average shells have a higher porosity and, as is shown in the figure below, as weight loss at 7 days of incubation increases then the age of embryonic mortality goes down.

In contrast to pheasant eggs there is no correlation between shell colour and pigmentation and ultrastructure in the red-legged partridge. It would be interesting to investigate whether shell colour can be correlated with weight loss and ultrastructure in other species with variable shell colour.

Relationship between % weight loss measured at 7 days and the age of embryonic mortality in blue pheasant eggs.

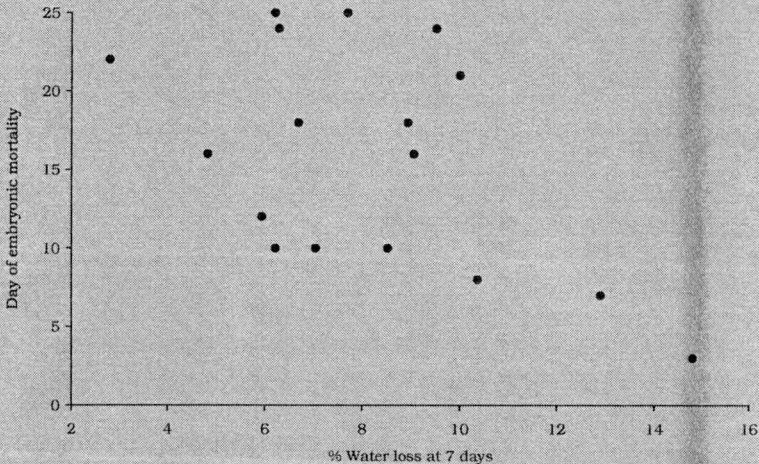

Figure 3.7. Scanning electron micrographs showing the variability of pore morphology in the eggshells of the Houbara bustard. Note the thin pore with a small funnel at the pore orifice in A and the wide pore in B. The funnel of the pore orifice extends deep into the palisade layer in C (arrow). Photographs by D. C. Deeming & G. K. Baggott. Scale bars = 50 μm.

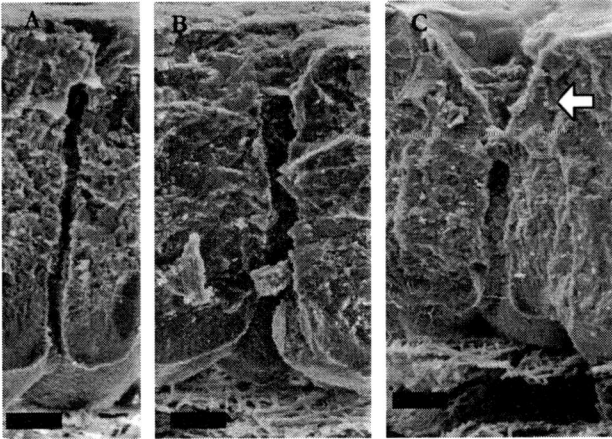

storks are unusual in that the pores open not to the outer surface of the shell but into an air-filled "honeycomb" structure overlying the palisade layer and with separate openings on the outer surface (Figure 3.4D). It remains unclear how these multi-branched pores, and the reticulated shell structures, are formed during shell deposition.

A key role of the pores is to allow respiratory gas exchange and it has been shown that pore number and dimensions scale with egg mass (Figure 3.8). That is, bigger eggs have more, larger diameter pores that allow more gas exchange proportionately dependent on the embryo's metabolism. Shell thickness also increases as egg mass increases (Figure 3.8).

Measuring shell porosity

The rate at which bird eggs lose weight during incubation is a function of the porosity of the egg and the humidity outside of the shell. Back in the early 1970s, Hermann Rahn, Amos Ar and Charles Paganelli began to analyse the physical process of weight loss from bird eggs and developed the idea of using water vapour conductance to measure the porosity of eggshells. In effect water vapour conductance put a figure on the combined effects of eggshell thickness, pore size and number in determining the porosity of the shell. All of their research, and all that it stimulated, has left us with a tremendous wealth of data on the water relations of bird eggs during incubation.

At a practical level weight loss is used as a measure of eggshell porosity under prevailing humidity conditions but this does not allow a full comparison between species and differing incubation environments. Water vapour conductance (G_{H_2O}) allows such comparisons and has proved a valuable tool to study incubation in nests and incubators alike.

To calculate G_{H_2O} the weight loss of an egg has be measured under known conditions of humidity and temperature. In essence:

Weight loss = water conductance x humidity difference across the eggshell.

Or in mathematical terms:

$$M_{H_2O} = G_{H_2O} \times (P_eH_2O - P_nH_2O).$$

Where: M_{H_2O} = daily water loss from an egg (mgH_2O/day); G_{H_2O} = water vapour conductance (mgH_2O/day/Torr); P_eH_2O = humidity inside the egg (Torr); and P_nH_2O = humidity outside the egg (Torr).

Now this equation may turn you off but stay with this – it is important in understanding the relationships between weight loss, humidity and shell porosity. Firstly, the humidity inside the shell (P_eH_2O) is always at 100% humidity,

Figure 3.8. Relationships between initial egg mass and total pore number per shell (black circles), shell thickness (μm, grey triangles) and pore radius (μm, white squares). Lines show the tend for each relationship.

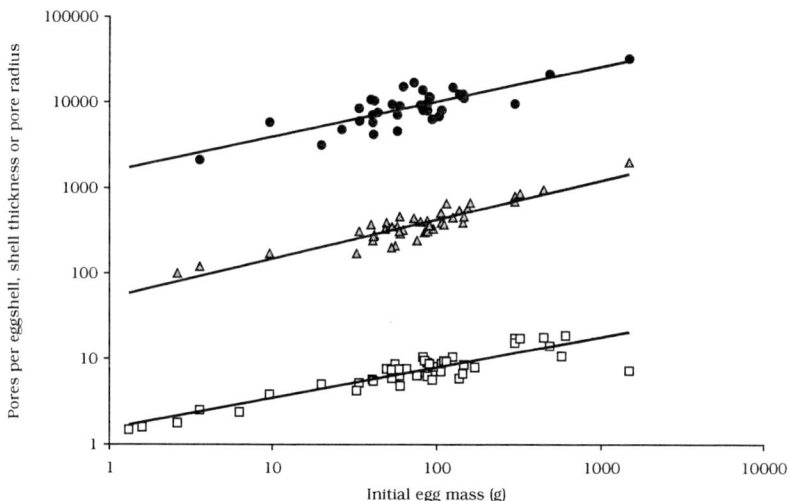

which at 37.5°C is 48.5 Torr. If the value of G_{H_2O} is fixed at laying this means that the humidity in the nest (P_{nH_2O}) is the controlling factor in determining weight loss.

So if you know the rate of weight loss and humidity in the nest then it is possible calculate water vapour conductance of the eggs being incubated. When determining G_{H_2O} eggs are often kept in dessicators with a drying agent to create a 0% humidity. Conversely, if you measure the weight loss of an egg with a known water vapour conductance then the humidity of the nest or incubator can be calculated. Weighing other eggs in the same nest allows their G_{H_2O} values to be calculated. This information can be very useful in understanding the optimal incubation environment in artificial incubators.

Water vapour conductance is a convenient way of expressing eggshell porosity. Conductance to oxygen and carbon dioxide can be determined in the same way but in practical terms it is easier to measure weight loss from the egg. Note that the loss of weight of an egg is due to water vapour alone because the weight of oxygen entering an egg is balanced by the weight of carbon dioxide leaving it. If G_{H_2O} is known that G_{O_2} and G_{CO_2} can be calculated.

As for other shell characteristics (Figure 3.8), both daily rate of weight loss and water vapour conductance scale with egg mass (Figure 3.9). This shows that almost all birds are exhibiting the same pattern of shell characteristics

Figure 3.9. Relationships between initial egg mass and water loss during incubation (mg/day; black circles) and water vapour conductance (mgH₂O/day/Torr; white squares). Lines show the tend for each relationship.

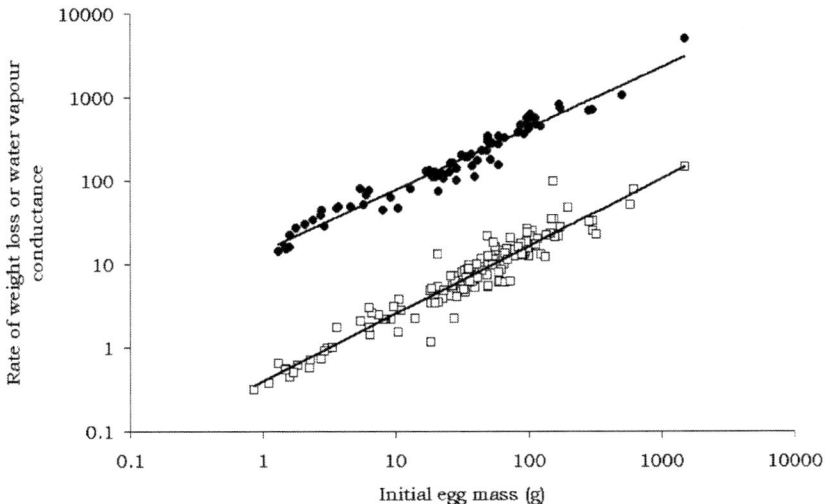

irrespective of egg size. The slope of the line for water vapour conductance reflect the importance of the role the pore plays in oxygen – carbon dioxide exchange. This aspect of embryonic development is described in more detail in Chapter 5 (see pp. 92–93).

Important aspects of egg composition

Although bird eggs are remarkably similar in their structure there are distinct differences in their gross composition as well in their chemical composition. Much of these differences are related to hatching maturity, *i.e.* whether the chick is precocial or altricial (see pp. 7–8). In this section I also describe the various functions of albumen and yolk before and during embryonic development.

Gross composition of albumen and yolk

Albumen has several roles in the egg. In the first instance its chemical composition and gelatinous structure act as significant anti-microbial agents. Any micro-organisms crossing the eggshell are either immobilised or killed by the albumen proteins (see pp. 39–40). Albumen is also a key reservoir of water and mineral ions, particularly sodium, necessary for normal development. Its proteins are also valuable for embryonic and post-hatching growth. Moreover, the albumen serves to act as a physical buffer for the yolk helping to absorb physical shocks and retain heat from the brood patch.

Yolk also has several roles. Primarily it serves as the nutritional source for the developing embryo. Rich in lipids and proteins, yolk provides the material needs of the embryo. Trace elements, such as vitamins and mineral ions, are also stored in the yolk. During development the yolk also serves as a waste store for bile salts. It also supplies energy and protein for the first few days of the bird's life after hatching.

A lot of research has been done on the relationships between the gross composition, *i.e.* the amounts of albumen and yolk, in the eggs of species producing hatchlings of different developmental modes. Hence, eggs of the super-precocial megapodes (see Box 5.2, p. 79) and the kiwis (Box 3.4, p. 37) have relatively little albumen and their yolk is energy rich (Table 3.2). This means that the water forms a relatively low percentage of the fresh egg contents. By contrast, eggs of precocial species have more albumen, smaller yolks and so have less energy; the higher albumen content increases the water content of the egg (Table 3.2). These trends are continued as the hatchlings become developmentally less mature with altricial species laying albumen and water rich but energy poor eggs (Table 3.2). On the whole this pattern also reflects the length of the incubation periods in these groups (see also pp. 62–64). Precocial species have energy rich eggs and long incubation periods com-

BOX 3.4 – THE REMARKABLE KIWI EGG

Restricted to New Zealand the Kiwis are remarkable birds in many respects. This certainly applies to their eggs and incubation behaviour.

Laying 1 or sometimes 2 large eggs (300–400 g) the female kiwi has to cope with production and laying of an egg that is 20–25% of her body mass (see illustration below [adapted after Calder, Scientific American, 1978] of an egg relative to skeletal size). This compares with values of ~15% of body mass seen in the smallest birds. The yolk forms around 65% of the contents making it the most energy-rich bird egg (over 12 kJ/g). This serves to support the embryo over a very long incubation period (over 70 days) and to produce a super-precocial hatchling, which like that of megapodes (Box 5.2, p. 78) is capable of fending for itself a few days after hatching.

Laid in a burrow the eggs are incubated by the male alone in most species although this system can vary. In mountainous areas the energy requirements of the male mean that he has to leave the egg for longer periods and so the female keeps the egg warm for short periods. The combined %attentiveness in these birds is equal to that observed in male-only incubation by individuals of the same species living by the warmer coast. The burrow is so small and the egg so large that turning is not possible in kiwis.

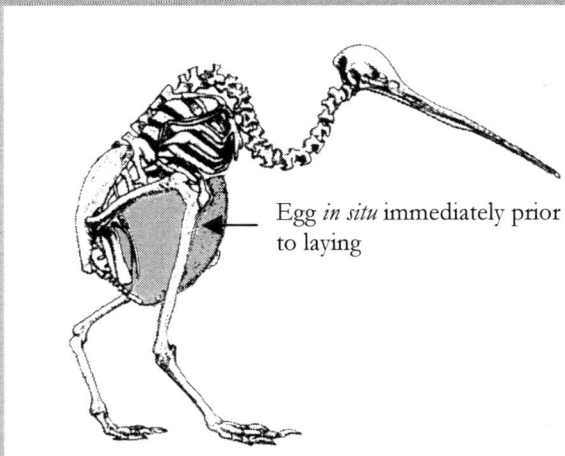

Egg *in situ* immediately prior to laying

Table 3.2. Albumen, energy and water composition of eggs and the water composition of hatchlings of different developmental modes.

Hatchling type	Albumen (%)	Energy (kJ/g)	Water - Fresh egg contents (%)	Water - hatchling (%)
Super-precocial	48.35	9.72	66.0	67.5
Precocial	63.34	7.43	73.4	72.9
Semi-precocial	68.05	6.84	76.9	77.9
Offshore and pelagic feeders	63.05	7.30	75.16	73.6
Semi-altricial	74.06	5.01	81.7	83.6
Altricial	77.35	4.57	83.6	83.9

pared with the short incubation periods of energy poor altricial eggs.

Exceptions to this pattern are those species of seabird that feed far off-shore and have to spend a lot of their time away from the nest. Compared with other semi-precocial seabird species that feed near to the shore, the eggs have less albumen but are more energy rich (Table 3.1). This is because the eggs of these species are often left unattended during incubation and so re-quire a higher energy content to survive the extended incubation period.

Another aspect of the provisioning of the egg with water is in the water content of the offspring at hatching. Altricial species are developmentally less advanced than precocial species and hence have a higher water content at hatching (Table 3.2). The water content of the egg at laying matches the water content of the hatchling because there is some metabolic production of water during development and the egg loses water during incubation (see Box 5.4, p. 90).

Chemical components of the egg contents

Albumen is composed of a wide variety of proteins and water. Proteins form almost all of the solid component of albumen and 10.5% of the total weight (Table 3.3). Water on the other hand forms 88.5% of whole albumen. Other components, such as carbohydrate, lipids and inorganic ions, are each only less than 1% of the total weight (Table 3.3).

There are numerous proteins in albumen, each having a specific role in protecting the egg from microbial attack (Table 3.4). These include proteins that prevent microbial growth by binding vitamins or iron that are important for bacterial metabolism. Similarly many proteins inhibit bacterial protease enzymes that would otherwise breakdown the egg contents. By contrast, lysozyme (also found in human saliva) is an enzyme that kills bacteria by destroying their cell walls.

The exact make-up of the albumen proteins is variable between species and this has proved useful in the study of the evolutionary relationships between different birds. For instance, ovotransferrin forms 12% of the albumen proteins in the fowl egg, compared with 10% in emu, cassowary and kiwi eggs and only 3% in ostrich and rhea eggs. Penguin eggs have their own specific form of ovalbumin. Whether these species differences reflect the potential microbial challenge that eggs could encounter has not been fully investigated.

The water and protein contents of yolk are much lower than in albumen and lipids feature more prominently (Table 3.3). Again carbohydrate and inorganic ions are minor components of yolk. Yolk also contains trace amounts of vitamins and minerals as well as carotenoid fat-soluble pigments that give its characteristic colour. Indeed, yolk colour reflects the amount and variety of pigments in the diet (natural or otherwise) rather than the quality of the egg.

The lipids in yolk come in a variety of different types. In the domestic fowl egg the yolk is formed of triacylglycerides (71%), phospholipids (21%), free cholesterol (6%), cholesterol ester (1%) and free fatty acids (1%). Just as the albumen proteins vary between species, research has shown that the lipid co-

Table 3.3. Gross chemical composition of the albumen and yolk of eggs of the domestic fowl.

Component	Albumen (% of weight)	Yolk (% of weight)
Water	88.5	47.5
Proteins	10.5	17.4
Lipids	0.02	33.0
Carbohydrates	0.5	0.2
Inorganic ions	0.5	1.1
Other compounds	-	0.8

Table 3.4. The protein components of albumen from the domestic fowl egg.

Protein	Percentage of total protein	Characteristics
Ovalbumin	54	Possible enzyme inhibitor and/or binds metals
Ovotransferrin	12	Binds iron
Ovomucoid	11	Protease inhibitor
Ovoglobulins	8	Foaming agents
Lysozyme	3.4	Bactericidal enzyme
Ovomucin	1.5	Virus inhibitor
Ovoinhibitor	1.5	Protease inhibitor
Ovoglycoprotein	1.0	?
Riboflavin-binding protein	0.8	Binds the vitamin riboflavin
Ovomacroglobulin	0.5	Protease inhibitor
Thiamin-binding protein		Binds the vitamin thiamin
Avidin	0.06	Binds the vitamin biotin
Cystatin	0.05	Protease inhibitor

Figure 3.10. Fatty acid profiles of total yolk from species of bird feeding on grain (black columns), vegetation (grey columns) and fish (white columns). Arrows of the appropriate colour highlight critical differences between the different diets.

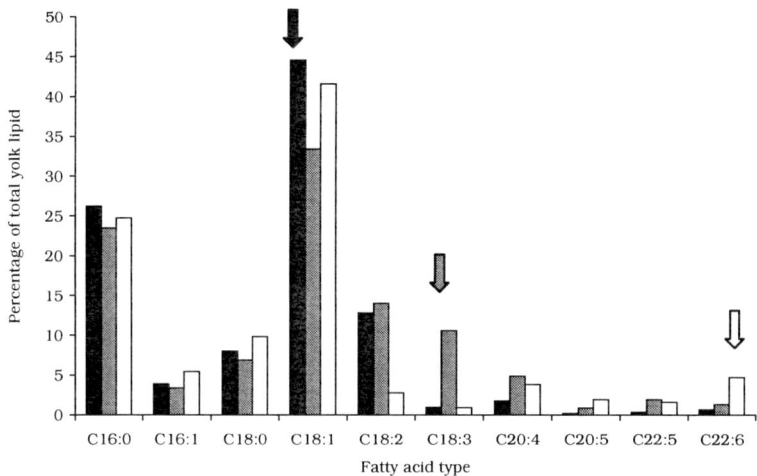

mponents of yolk vary between species and largely reflect the diet of the birds. The distribution of lipids between these component types in the fowl is exhibited by other precocial species but in a semi-precocial species, such as a gull, phospholipids are increased at the expense of triacylglycerides. The yolk of the altricial pigeon has only 58.0% triacylglycerides and 30.7% phospholipids. The water content of altricial yolk is also increased to ~65% of total mass. This all leads to a reduction in the energy content of the yolk and the egg as a whole (Table 3.2). This system of provisioning the egg with different levels of lipids is flexible and can be modified to reflect factors like the prolonged incubation period of offshore seabirds (Table 3.2). In semi-altricial penguin eggs the lipid profile is more similar to a precocial species but this is considered to reflect the long incubation period of these species.

Lipids are built up from fatty acids (long molecules of carbon atoms linked together in chains) derived from the diet of the bird. The composition of the fatty acids has been shown to vary between species and dependent upon their diet. Hence, species eating predominantly grains (*e.g.* poultry) have yolks rich in saturated and mono-unsaturated fatty acids (Figure 3.10). By contrast, birds feeding on vegetation have higher levels of the polyunsaturated fatty acids (Figure 3.10). Egg yolks of carnivorous species of birds (*e.g.* gulls and penguins) are comparatively rich in long-chain polyunsaturated fatty acids.

Finally, in several species it has been shown that the egg yolks from farmed birds are different in composition that yolks collected from wild birds. For the goose, pheasant and ostrich, yolks from farmed birds have very low levels of linolenic acid (1.3–2.7%) compared with high levels (17.6–27.9%) extracted from yolks from wild birds. It would appear that the diet of the farmed birds is producing yolks with atypical compositions but to date whether these differences are reflected in lower hatchability of farmed eggs is untested.

Summary

- Egg formation takes place in the ovary and the oviduct
- Bigger birds lay proportionately smaller eggs
- The yolk and albumen have a specific structure at laying
- The eggshell is made up of two fibrous membranes and a thick hard layer of calcium carbonate
- Pores permeate the eggshell and vary in morphology according to egg size
- Shell accessory material protects the egg contents from microbial contamination

- Shell characteristics like thickness, pore number, pore radius and water vapour conductance all scale to egg mass in a similar way
- Egg contents vary according to hatchling maturity
- Albumen plays a critical role in preventing microbial spoilage of the egg
- Yolk lipid content reflects the diet of the birds

4 - Fertilisation and Embryonic Development

Incubation provides the environmental conditions for embryonic development in birds. The processes that take the single cell, created by the fusion of an ovum and spermatozoon, through to a free-living bird are a wonder of nature but are rather complex. It is true to say that it is not necessary to know what is happening within an egg to be able to successfully incubate it. However, an appreciation of the changes that the embryo undergoes can assist in understanding how differing incubation conditions influence embryonic development. In this chapter I try to provide a brief introduction to avian embryology describing the principles of development. I have no intention of being too specific about what happens when and why but rather my approach is to present a broad picture of development.

Firstly I will deal with formation of the sex cells and fertilisation. The three phases of development are then described: 1) in the oviduct, 2) the differentiation phase during the first half of incubation, and 3) the growth phase during the second half of incubation. In this chapter I will also describe the pattern of extra embryonic development and how albumen and yolk are utilised by the embryo. I conclude the chapter by looking at the process of hatching, the differences in hatchling maturity and the factors affecting the incubation period.

Formation of sex cells

Reproduction in birds requires that the male and female sex cells join together to form a single cell – the "zygote", which then divides and eventually grows into the embryo. Sex cells are produced by meiosis (Box 4.1, p. 44) in the sexual organs: the testis in the male and the ovary in the female.

The testes in birds are located within the body cavity close to the kidneys and the spine. Each testis is linked to the cloaca (vent) by a *vas deferens*. The role of the testis is to produce spermatozoa. These are long thin cells packed with a nucleus, with half the normal number of chromosomes, in the "head" and a long "tail" with which it is able to swim (Figure 4.1). The spermatozoa are released and travel down the *vas deferens* and are mixed with seminal fluid before ejaculation as semen into the cloaca via the vent of the female during copulation.

Female sex cells are formed within the single left ovary again located adja-

44

BOX 4.1 – CELL DIVISION

Cells are basically bags of jelly containing small bodies, called organelles, that have specific functions within the cell. For instance, mitochondria produce energy and the nucleus contains a set of paired chromosomes (strands of DNA) that contain the blueprint for the working of the cell (and organism). In animals the cells are enclosed in a thin membrane.

Growth of cells occurs in two ways: an increase in size and an increase in number. Cells cannot grow in size indefinitely and so they split in half to form new "daughter" cells that can then grow in size. Embryos are formed by two cells combining to form a single cell which then grows in size and divides many times to eventually form a chick.

During basic cell division (mitosis) the cell simply splits into two equal daughter cells. As a prelude to division the chromosomes are copied to form two full sets, which then split into equal groups each with a full set of chromosomes. As they move apart the nucleus breaks down and the whole cell splits down the middle to form two smaller cells. Each has a new nucleus with a full set of paired chromosomes. This type of cell division is used for all cell types except sex cells.

Each cell has to have a full set of paired chromosomes and because an embryo is formed by the combination of two cells the sex cells have to have only half of the pair of chromosomes. Otherwise the new cell would have twice as much DNA as it needed. Formation of sex cells involves a special process of cell division (meiosis). Instead of the chromosomes splitting equally before cell division meiosis involves each chromosome splitting into two. One half of the pair goes into each daughter (sex) cell, which has only half of the chromosomes. Once two sex cells combine their nuclei combine and the number paired compliment of chromosomes is restored.

cent to the kidneys. The ovary is connected to the cloaca by an opened ended oviduct (see Figure 3.1 and pp. 21–22). Meiosis produces ova within the ovary and these become enlarged in turn forming the egg yolk described in Chapter 3 (see pp. 21–22). The actual sex part of the ovum is very small and comprises the pronucleus located at the top of the yolk in an area called the germinal disc. In effect the whole yolk is a single cell, which makes the ostrich yolk the biggest cell in the world!

Copulation and fertilisation

Copulation is the process of transferring sperm from the male bird into the female's reproductive tract. Mating in birds is often a precarious act. Most

Figure 4.1. Morphology of a typical avian spermatozoon.

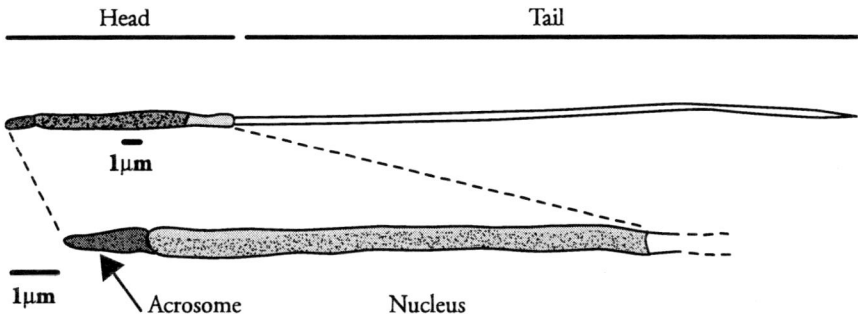

male birds lack any organ for inserting into female and otherwise bird anatomy does not readily allow the two vents to come together. Fortunately the two vents have only to be together for a second or two for transfer of semen (a mixture of sperm and a fluid that supports them). Some birds, notably ratites and waterfowl, do have a penis-like phallus that is inserted into the female vent to assist with copulation.

Fertilisation, being the fusion of the male and female sex cells, does not occur in the oviduct (as is the case in mammals) but rather has to happen within a few minutes of ovulation in the mouth of the infundibulum (see p. 22). Hence, the sperm have the monumental task of swimming up the oviduct to reach the yolk. In small birds this is only a few centimetres but in the ostrich the oviduct is over half a metre long. Furthermore, the movement of sperm is against the flow of any egg being formed and moving down the oviduct.

The female bird is able to cope with the vagaries of mating as well as not knowing exactly when the yolk will be released by storing sperm in the oviduct. Located at the bottom end of the duct next at the junction of the shell gland and the vagina (see Figure 3.1) the sperm storage tubules are blind-ended tubes extending into the oviduct wall. Here the sperm are nourished to maintain their viability and can be released at times many days after the previous copulation. There are also similar tubule-like structures at the top of the oviduct where the sperm can reside temporarily but it is generally considered that long-term storage does not occur here.

Fertilisation occurs between the time that the yolk is released from the oviduct and the time that the fibres of the outer peri-vitelline layer are laid down. After ovulation the sperm swim toward the germinal disc of the yolk and attach to the outer surface of the inner peri-vitelline layer. A secretion from the acrosome (Figure 4.1) dissolves away the fibres of the inner peri-vitelline layer to form a hole through which the nucleus can pass (Figure 4.2).

Many sperm make holes all over the inner peri-vitelline layer but there is a higher concentration of holes over the germinal disc. However, only one can provide the nucleus for fusion with the ovum's nucleus. The sperm have only 10–15 minutes before the outer membrane is complete thereby making the overall membrane too thick to penetrate. Many sperm get trapped by the deposition of fibres and can be seen under a microscope lying between the inner and outer peri-vitelline layers (Figure 4.2). More than 6 sperm holes in the inner peri-vitelline layer over the germinal disc is considered to be a good indication that the ovum has been fertilised.

The fusion of the male and female sex nuclei, which each have half the normal compliment of chromosomes, creates a zygote (with normal paired chromosomes) that then goes on to divide and form the embryo. This complex process can be simplified into three parts because there is considerable cell division in the oviduct, there is differentiation of different parts of the embryo into bodily organs and there is growth of the nearly fully formed bird up to the point of hatching. These processes are briefly described below.

Development within the oviduct

The avian zygote is a strange structure because in the first instance it consists of the whole yolk. The first cell division occurs by mitosis (Box 4.1, p. 44) some 5 hours after fertilisation and as the yolk moves down the oviduct more divisions take place. By the time the yolk reaches the shell gland there are eight cells cut off and lying above the yolk. Whilst in the shell gland cell division continues apace forming thousands of identical cells which come to

Figure 4.2. The sequence leading to fertilisation in birds: a) sperm swim towards and attach to the inner peri-vitelline layer covering the yolk; b) secretions from the acrosomes begin to dissolve away the fibres of the membrane; c) holes are made in the inner layer; and d) the nucleus of the sperm moves through the inner layer into the yolk in readiness for fusion with the female sex cell nucleus.

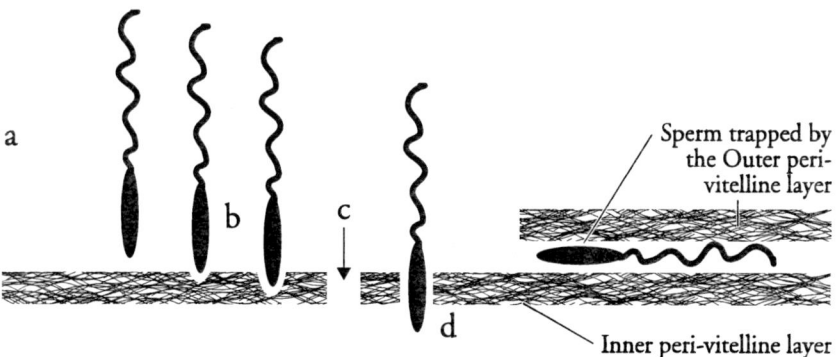

lie as a circular plate of cells. In cross section this blastoderm is only one cell thick making it relatively transparent (the *area pellucida*) except at its margins where there are more cells making it more opaque (*area opaca*). It is the cells of the *area pellucida* that eventually form the organs and tissues of the embryo.

At oviposition in poultry the embryo comprises 30,000–60,000 cells in the form of a circular "blastoderm" measuring 2–3 mm in diameter (even in the ostrich egg). Whilst the size of the blastoderm is unknown in many other species it is likely that all bird species are at approximately the same stage of development at egg laying. The exact stage of embryonic development is affected by the time in the oviduct and the time that eggs take to cool down after laying. Many embryos are at the gastrula stage of development, which is the start of the process of re-organisation of the cells to form the embryonic tissues and organs.

It is important to note that if fertilisation has not taken place then the germinal disc is able to divide and form a circular structure. This is the "blastodisc" of an infertile egg and is typically, an incomplete plate of cells resembling Swiss cheese when viewed under a microscope. In some turkeys a blastodisc is able to develop into embryos through a process of parthenogenesis (*i.e.* embryonic development without fertilisation).

Embryonic development during incubation

Embryonic development in birds has two key aspects. The first is the development of the embryo itself, which forms the body in readiness for growth to hatching. The second is the development of extra-embryonic components that are necessary to act as a life-support system for the embryo during its time in the egg. Bird development is characterised by being a closed affair: during incubation there is only exchange of heat, oxygen and carbon dioxide and loss of water. Indeed all of the water required for normal development is invested in the egg before the egg is laid. Therefore, the next two sections not only describe the changes in the shape and size of the embryo but also describe the growth of extra-embryonic membranes and the changes in the extra-embryonic fluid compartments, albumen and yolk.

Embryonic development is commonly describes in terms of "stages" which describe the sequence of key morphological stages (*e.g.* first appearance of eye pigment). Research has shown that these stages are essentially the same for most species investigated to date. Only the timing of each stage varies with respect to the incubation period; most stages take longer in larger species with long incubation periods. Systematic study of development has been studied in very few species and the domestic fowl is the best understood species. Hence, this species is used here as a good model for embryonic development.

48

Phase 1: Differentiation
 Embryonic development is strongly affected by the orientation of tissues. Whilst still in the oviduct the embryo develops a front and back, which are important in subsequent morphological changes. In birds the pattern of development proceeds more rapidly at the front end of the embryo. The left and right sides of the embryo are created by the formation of the primitive streak during a process called gastrulation.
 At the time of egg laying the blastoderm consists of apparently identical cells but they all have a specific role to play in subsequent development. Gastrulation involves cells moving around to begin to fold into new structures and cells types. In particular, endoderm, mesoderm and ectoderm cells are established (Box 4.2, p. 49). The end result is that the flat plate of cells is reorganised into 3-D structure. This has happened by day 3 in the domestic fowl egg.
 Early on in development the embryo and its extra-embryonic tissues are difficult to distinguish apart. The endoderm cells form the alimentary tract and associated organs as well as the yolk sac membrane and allantois. The ectoderm cells form the brain and nervous tissues and the skin as well as the amnion and chorion. Mesoderm cells line the spaces in between and contribute to muscle, the circulatory system and the extra-embryonic membranes. Gastrulation first moves cells around to form a bi-layer of cells and a "primitive streak". Then mesoderm cells migrate along the primitive streak to invade the space between the upper ectoderm and lower endoderm. Then this flat triple layer of cells begins to fold up to form other structures. For instance the brain and the heart develop from plates of cells that fold over to form tubes. In general, embryonic development proceeds at the head end first and progressively moves down towards the tail. By day 4 of incubation in the fowl the embryo has an obvious structure, with eyes, brain and heart clearly visible and the amnion and yolk sac membrane are well developed (Figure 4.3).

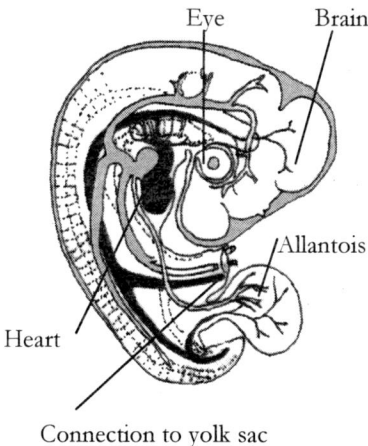

Figure 4.3. Appearance of a 4 day-old fowl embryo.

Eye   Brain

Allantois

Heart

Connection to yolk sac

Table 4.1 gives some indication of the periods of development of some of the major bodily tissues in the domestic fowl embryo. Many tissues are fully formed early in development

---

BOX 4.2 – TYPES AND ROLES OF EMBRYONIC TISSUES

All of the tissues of the body are derived from three types of cell: ectoderm, endoderm and mesoderm that develop from "stem cells" that are some the first to form during early cell division of the embryo and for a period can develop into any cell type. However, once the fate of each cell to produce a certain tissue types is established then it is unable to switch to another.

Ectoderm is found in the outer layers of the body and forms the skin. Nervous tissues are also formed from ectoderm cells that fold over to form a tube early in development.

Endoderm cells form the innermost layer of the embryo and line the alimentary tract and organs like the lungs. These cells are thin and are easily traversed by small molecules, such as digestion products.

Mesoderm cells provide a variety of tissue types that form the bulk of the bird. Vascular mesoderm is able to produce blood vessels but avascular mesoderm forms other organs rather than the circulatory system. Tissues like bone and muscle are produced from mesoderm cells.

---

but many others take much of the incubation period to be completed.

By day 7 of incubation in the domestic fowl embryo the egg contents have dramatically changed from that seen when the egg was laid (compare Figures 3.3 and 4.4). The embryo has developed to an extent that it is beginning to be recognisable as a bird and there is extensive growth of the extra-embryonic membranes. There is a considerable amount of sub-embryonic fluid and a smaller volume of thicker albumen (see p. 56). These changes continue to proceed the embryo gets bigger and more recognisable as a bird and the extra-embryonic membranes grow to line the yolk, albumen, embryo and inner surface of the shell membranes. Therefore by 12 days of incubation in the domestic fowl the embryo is a small bird surrounded by a complex of extra-embryonic membranes (Figure 4.6). Around 95% of the incubation period and the embryo now enters a phase of growth prior to hatching.

Phase 2: Growth

Although the embryo has gone from a small circle of cells to a recognisable bird by the end of the differentiation phase in most cases it is still very small and is incapable of hatching. Therefore, embryos for the species illustrated in Figure 4.5 weigh only 1–2 g at 50% of their respective incubation periods. During the second phase of development, which usually coincides

Table 4.1. Approximate periods of main development of selected tissues in the domestic fowl embryo.

Organ or tissue	Main period of development
Brain and spinal column	12 hours – 3 days
Eye	2–8 days
Heart and circulation system	2–5 days
Digestive system	1–7 days
Respiratory system	4–13 days
Muscles	2–12 days
Skeleton	5–21 days
Limbs	3–8 days
Kidneys	4–18 days
Gonads	3–8 days
Feathers	7–20 days
Yolk sac membrane	2–12 days
Chorion and amnion	1–3 days
Allantois	3–12 days

with the second half of incubation, the embryo grows to all but fill the eggshell. How it "knows" how big the eggshell actually is remains unclear but it seems that the embryo has to achieve a critical size before it can hatch.

Growth simply involves the embryo increasing its size by mainly increasing the number of the cells in each organ. The rate of increase in weight depends on the size of the egg, because it dictates the eventual size of the hatchling, and the length of incubation (Figure 4.6). Hence an embryo in a big egg with a long incubation period will grow relatively slowly.

A critical feature of avian development is the difference between precocial and altricial species. When compared in relative terms (Figure 4.7) these groups exhibit different patterns of growth with precocial species increasing in size more quickly than altricial species during the first half of incubation. During the second half of incubation the altricial species grow faster. How-

Figure 4.4. Diagrammatic representation of the contents of a domestic fowl egg at around 7 days of incubation. A) in cross-section and B) the part next to embryo indicated by rectangle.

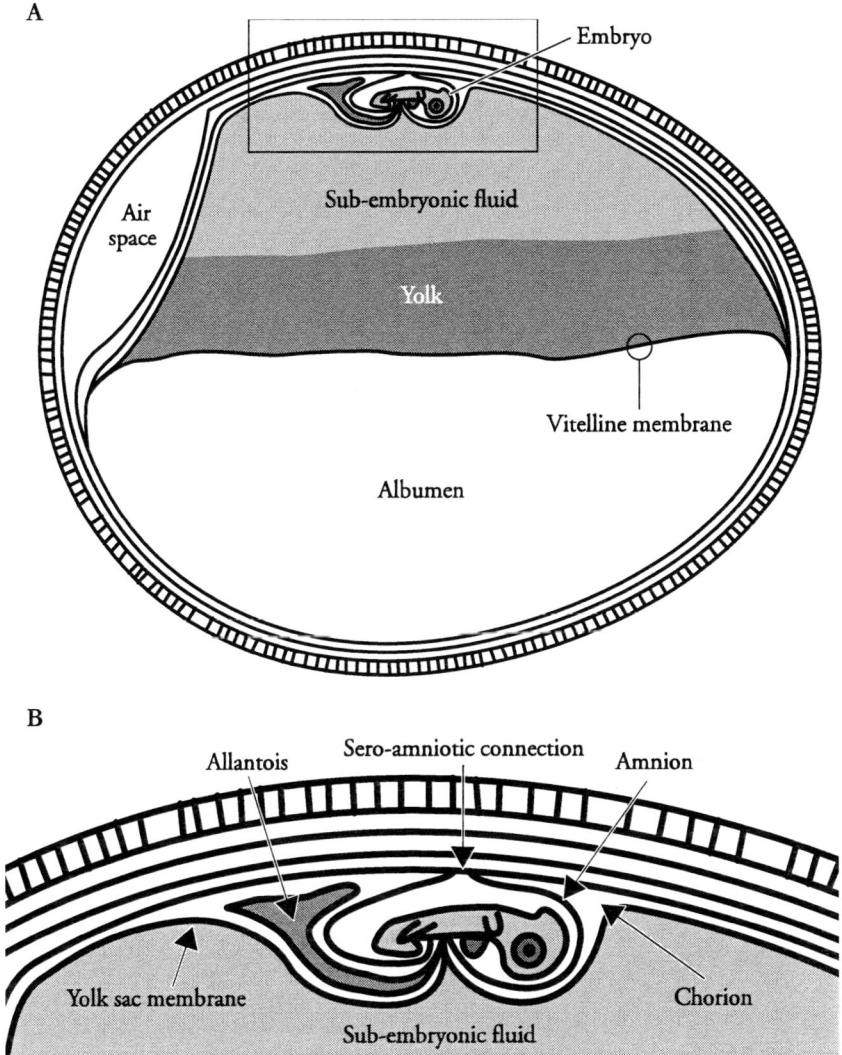

A

B

ever, as was shown in Table 3.2 altricial hatchlings have a higher water content than in precocial species. During development the water content of the embryo changes from being around 92–93% on day 7 in the domestic fowl but this drops to 73–75% in the hatchling. By contrast the water content of an altricial hatchling averages around 83%. This means that an altricial embryos at a time equivalent to day 16 of incubation in the fowl embryo. For this reason it is interesting to correct the length of incubation to the 76% of the incubation period of precocial species. The result is that embryonic growth during the first half of incubation is the same in altricial and precocial species but after 50% of the incubation period the altricial species grow very much faster (Figure 4.8).

There is a limited amount of organ development during the second half of incubation. For instance, the feathers mature in the embryos of those species which have more mature hatchlings species. The embryos also prepare for hatching during the latter stages of the second half of incubation (see pp. 57–63).

Extra-embryonic development

The egg is a capsule for the embryo to develop in isolation from the outside world. Only heat and gases traverse the shell and so the embryo has to be

Figure 4.5. Diagrammatic representation of the organisation of membranes of a fowl embryo within an egg after 12 days (55%) of incubation.

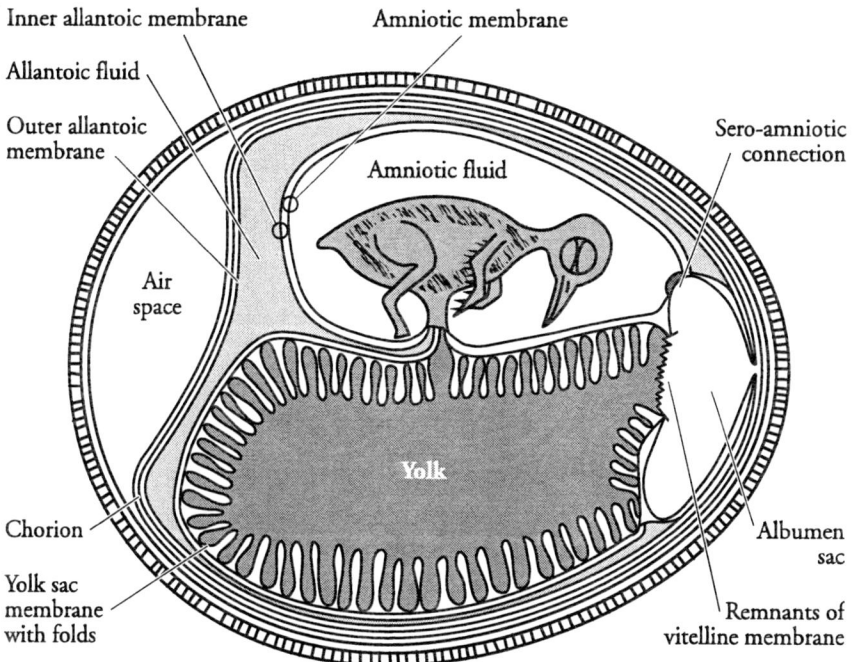

Inner allantoic membrane   Amniotic membrane

Allantoic fluid

Outer allantoic membrane

Amniotic fluid

Sero-amniotic connection

Air space

Chorion

Yolk sac membrane with folds

Yolk

Albumen sac

Remnants of vitelline membrane

Figure 4.6. Size of embryos of four types of embryos during incubation: the crow and pigeon produce altricial hatchlings and the pheasant and quail produce precocial offspring.

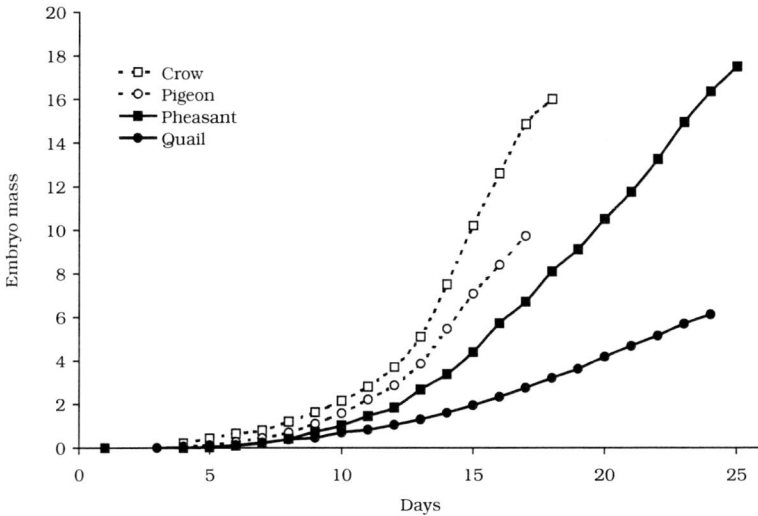

provisioned with all of its nutrients and water when the egg is laid. The embryo is reliant on extra-embryonic membranes in order to utilise these nutrients. These membranes are extensions of the embryo's body and are critical for providing the ideal environment for development. An understanding of the role of these tissues is crucial in comprehending how embryos develop.

The amnion is an extension of the body wall of the embryo. Comprised of ectoderm and avascular mesoderm (Box 4.2, p. 49) the amnion is a muscular layer of only a few cells thick that grows up over the head and tail of the embryo during the first 2–3 days of incubation. These two folds of membrane meet at the sero-amniotic connection to form a fluid filled compartment (Figure 4.3). This serves to protect the embryo from physical shocks experienced by the egg. As the embryo grows the amniotic sac increases in volume. During the second half of incubation the sero-amniotic connection ruptures (Figures 4.5 and 4.9) allowing albumen proteins to move through into the amniotic fluid where they are swallowed by the embryo (see p. 56).

The chorion is continuous with the amnion and consists of the same tissues (Figure 4.4). It forms the "outer skin" of the embryonic tissues as a whole. The primary role of the chorion is to interface between the inner shell membrane and the blood vessel rich membranes that exchange the respiratory gases (oxygen and carbon dioxide). This is the yolk sac membrane for the first

Figure 4.7. Size of embryo of four types of bird expressed as a percentage of the incubation period and the hatchling mass: the crow and pigeon produce altricial hatchlings and the pheasant and quail produce precocial offspring.

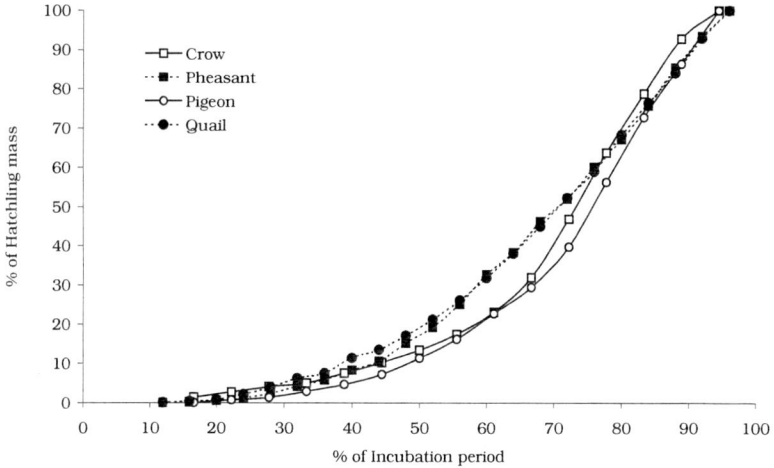

Figure 4.8. Pattern of growth of altricial (white symbols) and precocial (black symbols) species. Note that the incubation period and hatchling mass are expressed as a total so as to aid comparison and the length of the incubation period is corrected to match the water content of the chick at hatching.

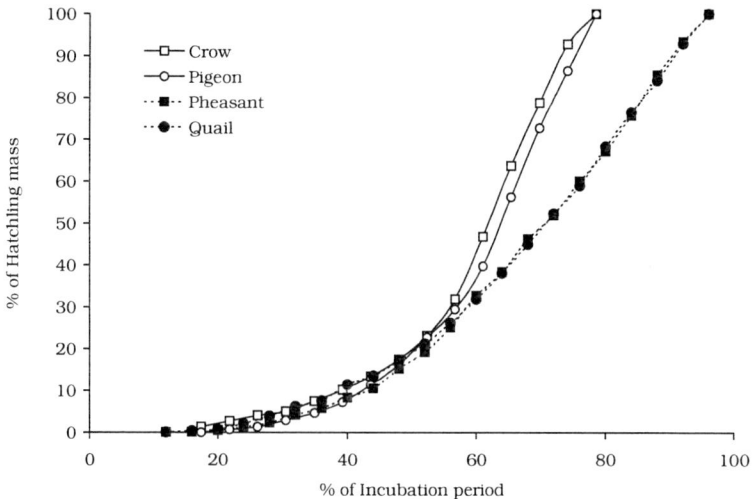

Figure 4.9. Scanning electron micrograph of a cast of the sero-amniotic connection (with the membranes removed) on day 14 of incubation in the domestic fowl. Albumen proteins move through the holes indicated.

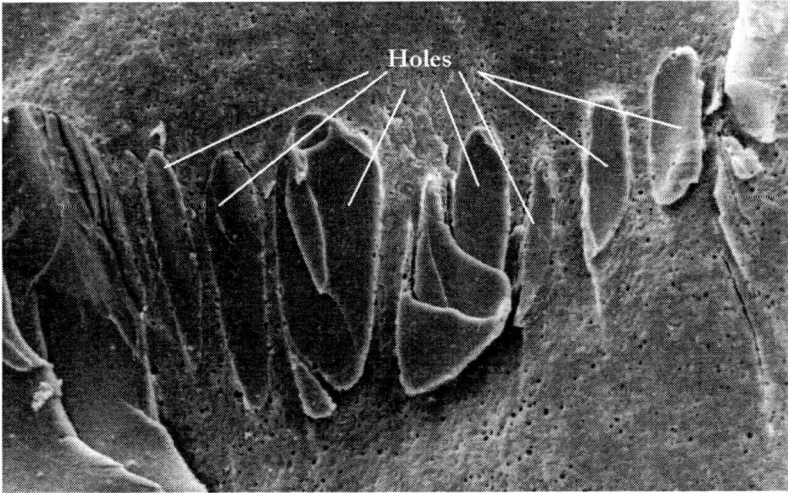

third of incubation and the outer allantois for the rest of development.

The yolk sac membrane is continuous with the small intestine of the embryo and is comprised of endoderm and vascular mesoderm (Box 4.2, p. 49). It forms within a couple of days of the start of incubation and it eventually completely encloses the yolk. Its primary role is in the breakdown and uptake of yolk, for which reason it develops finger-like projections into the yolk that increase its surface area (Figure 4.5). The yolk sac membrane is also critical in the formation of sub-embryonic fluid (see p. 56) and is the first major respiratory membrane.

The allantois is an outgrowth of the hind gut visible around day 4–5 of incubation in the fowl embryo and, like the yolk sac membrane, is made up of endoderm and vascular mesoderm (Box 4.2, p. 49). It forms a fluid filled sac that serves to act as a reservoir for urine produced by the embryonic kidney but also acts as a general water reservoir for the embryo. The outer part of the membrane combines with the chorion (Figure 4.3) and as the chorio-allantoic membrane it comes to line the entire surface of the inner shell membrane (Figure 4.4) where it serves as the primary respiratory membrane for the embryo during the latter two-thirds of incubation. During the last third of incubation the chorio-allantois also secretes a weak acid that dissolves away the inner surface of the calcitic eggshell thereby supplying calcium ions for bone formation. The inner allantois assists in yolk sac retraction (see pp. 58–59).

56

These membranes assist the embryo in the utilisation of albumen and yolk during incubation. These form almost 100% of the contents of the egg at ovi-position and are absent in the egg after hatching.

Albumen is utilised in two phases. The first phase involves the transfer of sodium ions from the albumen to the yolk by the yolk sac membrane. During the first third of incubation water and chloride ions move as well and a salty solution, the sub-embryonic fluid, forms on top of the yolk and below the yolk sac membrane and the embryo (Figures 4.4 & 4.10). The two liquids re-main relatively unmixed because the fat-rich yolk finds it hard to mix with the water (just like oil and vinegar salad dressing that remains in separate layers unless shaken). This process ensures that the embryo has access to sufficient water during later stages of development. The result is that the albumen's structure is lost (Figure 4.4) and its volume decreases considerably (Figures 4.5 and 4.10). Sub-embryonic fluid reaches a maximum volume around 7days of incubation in the fowl embryo and declines thereafter (Figure 4.10). The water is essentially moved around the egg by the embryo and goes to form allantoic fluid and amniotic fluid, which increase in volume as development proceeds (Figure 4.10). Only towards the end of development do these fluid compart-ments dry out in preparation for hatching.

Only water and ions are removed from the albumen and so the remaining proteins become more concentrated. They are eventually surrounded by the chorio-allantois to form the albumen sac (Figure 4.4) but they remain next to the remnants of the vitelline membrane and the sero-amniotic connection (Figure 4.4). From day 12 in the fowl egg, the membrane forming the sero-amniotic connection breakdown to form a series of holes between the albu-men and the amniotic fluid (Figures 4.5 and 4.9). For a reason not yet fully understood the albumen proteins move through the holes and dissolve in the amniotic fluid. In turn the amniotic fluid is swallowed by the embryo and so the albumen proteins move unchanged into the embryonic gut. Some are bro-ken down and utilised by the embryo but most simply enter the yolk sac via the yolk stalk that connects the yolk sac to the small intestine. The end result is that all of the albumen eventually disappears (Figure 4.10) and the protein content of the yolk increases during the second half of incubation.

The composition of the yolk changes with time. The influx of water to form the sub-embryonic fluid means that the water content of the yolk in-creases slightly during this time. Although the protein content may increase the lipid content of the yolk declines as the embryo uses it for its primary source of energy. Yolk also acts as a store of various waste products, primarily bile salts produced by the liver. These changes mean that the total mass of yolk hardly changes for most of the incubation period (Figure 4.10) and there is a considerable amount of yolk left to be retracted into the body cavity

Figure 4.10. Changes in the volume or mass of the various extra-embryonic fluid compartments and the embryo during development in the domestic fowl.

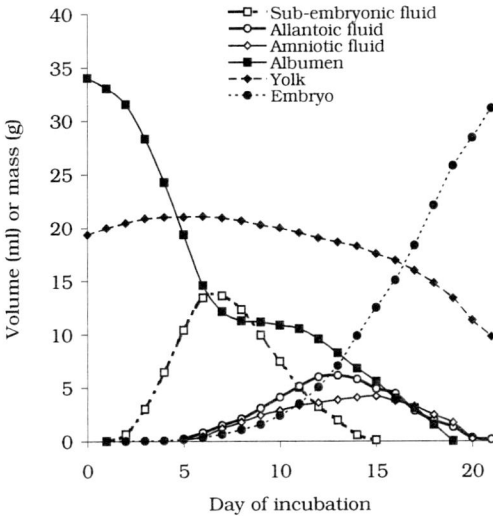

before hatching (see pp. 58–59).

The other significant change in the egg contents is the formation of an air space (Figure 4.5; Box 4.3, p. 58). This is the result of the loss of water vapour from the egg and the decrease in the volume of the egg contents being replaced by air diffusing in through the shell pores.

Hatching

Hatching is a critical period for the developing embryo because it marks the transition between life within the confines of the egg and its life as a free-living bird. Although hatching itself involves successfully breaking the shell the embryo undergoes a period of preparation for the hatching process during the later stages of incubation.

Assuming that embryonic development has gone according to plan and the embryo is the correct size and stage of development for hatching then it will embark upon a sequence of events that will lead to it leaving the egg. In most species the embryo has to initially orientate itself within the egg bringing its beak in line with the air space (Box 4.3, p. 58). This process starts several days before the hatching process and involves the head moving round so that it lies under the right wing and the tip of the beak is pressed against the inner shell membrane delimiting the air space (see Box 9.3, p. 180).

At the same time the extra-embryonic fluid compartments are draining of fluid. The allantoic sac is first to dry out followed by the amniotic sac. These fluids are simply absorbed by the embryo. If weight loss from the egg has been excessive then the allantoic sac may dry out several days before hatching begins. The uric acid (the main waste product in bird urine) dissolved in the allantoic fluid precipitates out and forms the white solid deposits often seen within hatched eggshells. Once the amniotic sac is dry then the amnion over the embryo degenerates.

Yolk sac retraction

As hatching time approaches the embryo begins to withdraw its yolk sac so as to bring into the body a huge reserve of nutrients necessary for the first few days post-hatching until the bird has learnt to feed or is being fed by its parents. Furthermore, the yolk sac contains waste products, such as bile that gives it a bright green colour, that the embryo was unable to dispose of whilst in the egg. The hatchling is able to reprocess the waste material and excrete it in the normal manner.

BOX 4.3 – THE AIR SPACE

Loss of water vapour from the egg during incubation leads to the formation of an air space within the shell. This forms between the inner and outer shell membranes typically at one end of the egg (Figure 4.5). In asymmetrical eggs the air space forms at the blunt end of the shell and can be used to define the "blunt end" in those eggs with a more symmetrical shape.

The air space is filled with air that diffuses in across the pores forming a bubble between the shell membranes. The exact composition of the air more reflects the levels of the gases dissolved in the embryo's blood. Therefore, at the end of incubation the air space is low in oxygen and high in carbon dioxide.

Bubbles that form within the egg contents and under the inner shell membrane appear to be highly disruptive to normal embryonic development and those eggs with "floating" air spaces usually fail to develop.

The air space has two main roles: 1) to provide an asymmetry in the egg that causes the egg to lie unevenly and allows the embryo to locate the position of the air space (see pp. 94–95). 2) to provide air for the embryo's first breath after internal pipping (see p. 59) which is used to expand the air sacs and lungs prior to external pipping and hatching. Embryos developing with their beaks away from the air space find it much harder to hatch (Box 9.3, p. 177).

Yolk sac retraction involves bringing the yolk sac into the abdominal cavity by the contraction of the membranes surrounding the yolk sac. Yolk is surrounded by the yolk sac membrane, which is connected to the small intestine by the yolk stalk. Outside of that the yolk sac is enclosed by a membranous "bag" consisting of the amnion where the yolk is closest to the embryo and by the inner allantoic membrane for the rest of the area away from the body (Figure 4.11). The yolk sac is pushed into the body by the muscles contracting in the amnion and inner allantois thereby making the volume of the "bag" surrounding the yolk sac smaller. The process takes place over a couple of days and starts with the disintegration of the amnion and inner allantois lying over the embryo's body. The amnion over the yolk sac is maintained. Initially contractions in the muscles in the inner allantois begin to reduce its area and hence push the yolk sac through the hole delimited by the amniotic ring into the body cavity (Figure 4.11A). Once the inner allantois has reduced as far as it can go, the ring of amnion attached to the inner allantois begins to contract further decreasing the volume of the "bag" outside of the body (Figure 4.11B). Once this ring has closed then the amnion covering the yolk sac contracts pulling the ring in towards the body wall (Figure 4.11C). The amnion finally folds to form the ring of muscular tissue clearly seen surrounding the navel and effectively seals the yolk inside the body wall (Figure 4.11D).

The hatching sequence

The process of hatching follows a series of events that are the essentially same for most species. At internal pipping the beak is pushed through the inner shell membrane into the air space thereby providing the chick with its first breath of air (see Box 4.3, p. 58). Many species have an "egg tooth" on the end of their beaks that, despite not being a tooth, is sharp enough to cut the membrane as the embryo pushes its beak upwards. The ostrich does not pip in this way but rather uses its beak, which has an especially padded tip, to rub away at the eggshell membranes at a position well away from the air space. Internal pipping then occurs by the embryo pushing up with its neck to pull the inner shell membrane over the beak.

Once in the air space the bird uses the air to fill its lungs and air sacs in preparation for normal lung breathing. This process is critical because it will provide the bird with sufficient oxygen during the later stage of incubation. After what can be a considerable amount of time the chick is stimulated to break the shell (external pipping) and gains access to fresh air for the first time. Again there can be a delay after external pipping because the embryo has to reduce the levels of carbon dioxide in its blood and increase its oxygen levels. Once this has been achieved the chick rotates around within the shell chipping away at the shell so that it can break a hole big enough for it to push

Figure 4.11. Stages in the process of yolk sac retraction. A) Initial stage of yolk sac retraction – the amnion and inner allantois disintegrate above points a and b. The inner allantois around the yolk sac contracts (curved arrows) and begins to push the yolk into the abdominal cavity. B) The inner allantois has contracted as much as it can and the amnion begins to contract bring points a and b together (arrows). C) Once points a and b are adjacent the amnion over the yolk sac begins to contract and shorten (arrows). D) Yolk sac retraction completed with navel ring composed of amniotic membrane.

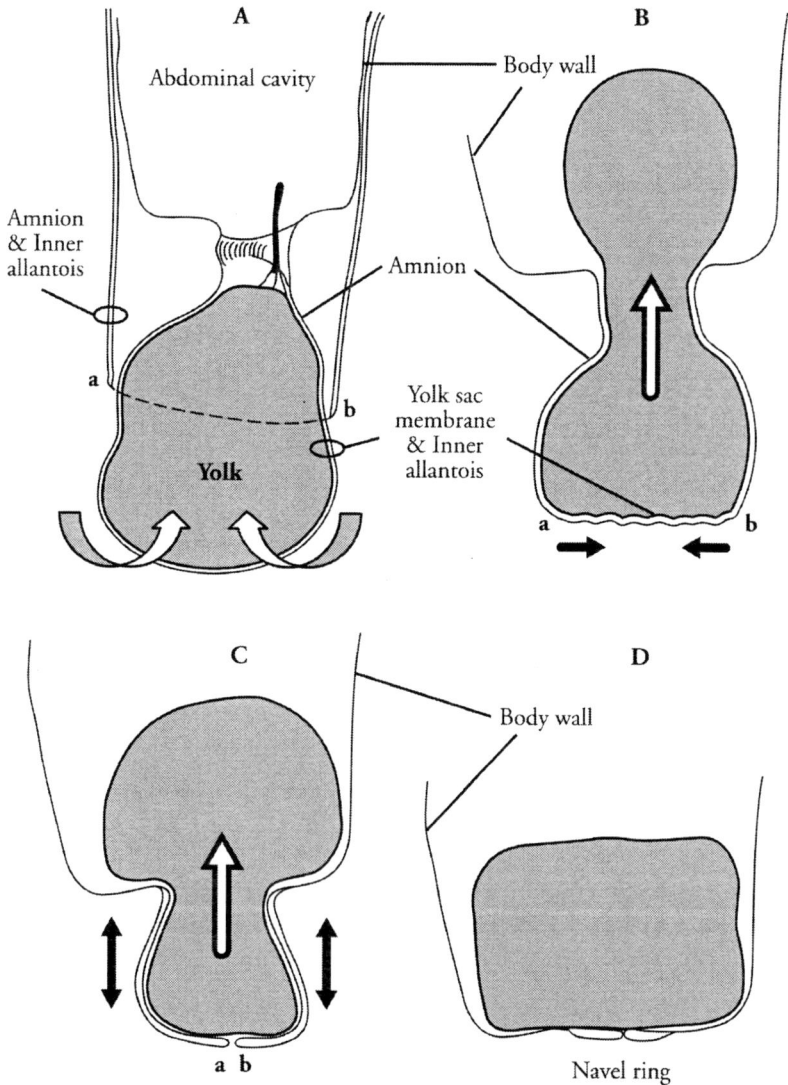

the shell off by flexing its body and being then free to kick its way out of the shell.

The process of breaking the eggshell varies between species and has been categorised into two groups based on the appearance of the eggshell after the chick has left (Figure 4.12). Many species follow the "symmetrical" pattern of hatching characterised by the upper part of the eggshell being cut away by the chick as its rotates within the shell. The end result is that the top half of the eggshell is cut away allowing it be pushed off leaving the remaining shell with an even appearance (Figure 4.12A). The degree of rotation during hatching varies between species. In general (and there are exceptions) the degree of rotation is smaller in precocial species with bigger eggs. Ratite chicks tend to rotate less than 90°, whereas swans tend to rotate between 180–200°. Geese chicks rotate between 200–220° compared with 240–300° in species of small duck. Quail species rotate between 300–360°. Altricial species rotate between 240–360° and with the top of the shell being cut off completely in the smaller eggs.

By contrast, many shorebirds, especially those with long beaks, make oblique pip holes, fail to cut off a neat part of the upper part of the shell and after breaking the shell emerge from the side of the egg. The result is an eggshell that looks unbalanced (Figure 4.12B). The American woodcock makes a pip hole near the blunt end of the egg and then uses its beak to hold on to the shell whilst it levers its way out of the shell with its neck.

This process of hatching does not happen in megapode chicks (Box 5.2, p. 79). During hatching in these species the beak is pointing towards the narrow pole of the egg and the chick moves around so vigorously within the shell that

Figure 4.12. The appearance of hatched eggshells of A) the symmetrical hatching pattern of the black-headed plover and B) the asymmetrical hatching pattern of the lapwing. White areas are exposed shell membranes.

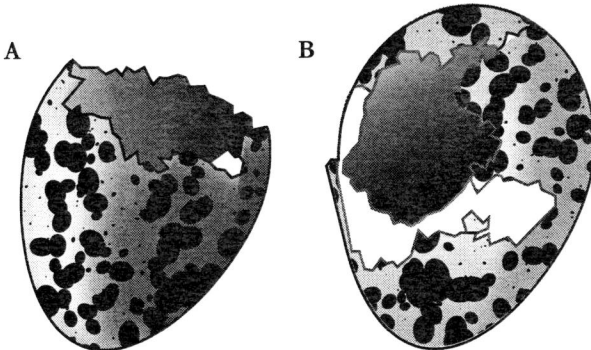

BOX 4.4 – EMBRYONIC COMMUNICATION WITH THE OUTSIDE WORLD

Recent research has shown that embryos may use nitric oxide (NO) gas to communicate with the incubating bird sitting on the egg. The brood patch has receptors which sense the amount of NO production by the egg and experiments have demonstrated that factors such as cold or hypoxia can affect the gas levels. Whether these changes elicit any response from the adult has yet to be demonstrated.

Once the embryo has filled its lungs and air sacs after internal pipping it is able to produce clicking noises (around 1–2 clicks per second) from its syrinx (the avian "voice box"). These clicks occur in altricial species and in several precocial species they assist in synchronising hatching of the clutch. Many precocial species can also vocalise during hatching.

Calling during hatching can also be heard by the incubating adults and may help in parent–offspring recognition or in telling the parents that hatching is imminent. In the American pelican vocalisations have been shown to encourage the adult to tend the egg. Vocalisation rate is higher in cool eggs and it appears to stimulate incubation behaviour in the adults. Calling by the embryos may help in ensuring that the nest is not prematurely abandoned.

---

it cracks the calcitic layer with its claws and destroys it. The chick then rotates cutting through the shell membranes before emerging. The megapodes are also unusual in that they are unable to lose sufficient water during incubation and are left with a lot of allantoic fluid that has to drain away before the chick can emerge. The chick then has to dig through the rotting vegetation or soil to make its way out of the incubation mound or burrow.

Control of hatching

The processes of hatching are largely under control of hormones produced by the thyroid gland. The pattern of hormone secretion seems to be linked more closely with real time rather than developmental stage. In some cases when, for instance the embryo has not been growing fast enough and is too small, the hormones set the hatching process in motion even though the embryo is not in a fit state to hatch. In these cases you see small, dead in shell embryos that tried to absorb huge yolk sacs and may have even internally pipped before dying.

The gaseous environment within the eggshell plays a critical role in stimulating hatching. High levels of carbon dioxide and low levels of oxygen in the air space are very potent at stimulating external pipping although the former is more potent. During hatching embryos are very tolerant to these abnormal levels of gases. Fowl embryos will tolerate 7.5% carbon dioxide in the air outside of the egg without any ill effects.

The behaviour of the adult is very important during hatching because the incubation environment has to be maintained to prevent chilling of the embryo and excessive weight loss from the shell membranes that could dry up and trap the embryo in the shell. Therefore in many species, there is vocal communication between the embryos and the adult. In many precocial species exhibiting hatching synchrony there is also embryo–embryo communication in order to synchronise hatching (See Box 4.4, p. 62).

Hatchling size

Under ideal circumstances the mass of the hatchling will be an optimum proportion of the initial mass of the egg. Therefore, if the humidity conditions have been ideal and the egg has lost the correct amount of water then the embryo will have a water content equivalent to the egg contents at lay. For precocial species this values is ~69% of the initial egg mass. The remainder is made of weight loss during incubation (15% of initial egg mass), weight loss during hatching (5%), shell mass (10%) and embryo meconium (1%). For altricial species the value is ~73.5% and probably reflects a lower rate of water loss (2.5%) during the shorter hatching period and relatively smaller eggshell mass (8%).

The incubation period

Embryonic development is defined by the period from the initiation of incubation (see pp. 66–67) up to hatching, after which the bird is a free-living organism. As has been shown for other parameters of egg biology (*e.g.* eggshell parameters [Figure 3.8]) the incubation period (I_p) of birds scales with the size of the eggs (Figure 4.13). Therefore, small eggs tend to have short incubation periods whereas large eggs have long incubation periods. There seems to be a minimum incubation period of 10–11 days, which applies to species with egg sizes ranging from 0.5 to 1 g. At the other extreme, ratite eggs have long incubation periods but these are variable. Despite having eggs of approximately the same size the emu incubates for ~56 days compared with ~39 days in the rhea. Most eggs lie in between the extremes of egg size and I_p.

For any given egg size it is most common to see an extended incubation period rather than one which is relatively short (note the points well above the line in Figure 4.13). Longer than normal incubation periods are associated with eggs that are not turned (Box 3.4, p. 37 & Box 5.2, p. 79) or with species

---

BOX 4.5 – INTERRUPTED INCUBATION AND EGG CHILLING

Incubation is far from continuous in many species of bird but the periods spent off the nest are relatively short typically measuring only minutes (see pp. 71–75). However, in albatrosses, shearwaters and petrels (Procelliformes) incubation is regularly interrupted. These species share incubation duties, which allows the non-incubating parent to fly out to sea to feed often being away for several days. Unfortunately the incubating parent is not always relieved before its own hunger drives it off the nest to feed. The end result is an egg left to chill for hours or days at a time.

For most species of bird this would spell disaster for the embryo but in the Procelliformes the embryo has developed considerable resistance to chilling. Embryos have been recorded as remaining viable after 10–12 days at a nest temperature of 10°C. The main effect of egg chilling is a prolonged incubation period. This system of incubation is so developed in the group that the egg contents and shell have been adapted to the prolonged incubation period (see pp. 63–65). The embryo has modified its metabolism and rate of growth. For instance, the egg of the wedge-tailed shearwater has an incubation period over twice the length of the fowl egg, which is the same mass. Relative to the fowl, patterns of oxygen consumption and embryonic growth are slow in this species. There has also been a reduction in the water vapour conductance of the eggshell, which prevents excessive weight loss during the long incubation period.

The primary advantage of interrupted incubation lies in the ability of both parents to feed for sufficient times and to counter the unpredictable environment that the animals encounter. Both parents are required for chick rearing and so body condition needs to be maintained.

---

that experience egg-neglect and subsequent chilling (Box 4.5, p. 64). For its size the ostrich may have the largest eggs but their incubation period is relatively short. It is possible that embryos of brood parasites (Box 5.1, p. 69) have accelerated development so as to shorten the incubation period and give the hatchlings a head start over the host's eggs.

It is important to note that incubation periods are not fixed and can vary according to incubation conditions. Therefore, in any clutch eggs can hatch out over a 2–3 days period. In most cases the hatchling will emerge once its development has been completed but in some precocial species there is considerable communication between embryos in the run up to hatch (Box 4.4, p. 62) and this can sometimes retard hatching in advanced chicks or to accelerate development in late embryos. The end result is that hatching is synchronised.

65

Figure 4.13. Relationship between initial egg mass (g) and length of the incubation period (days) for a variety of different species from hummingbirds to the ostrich.

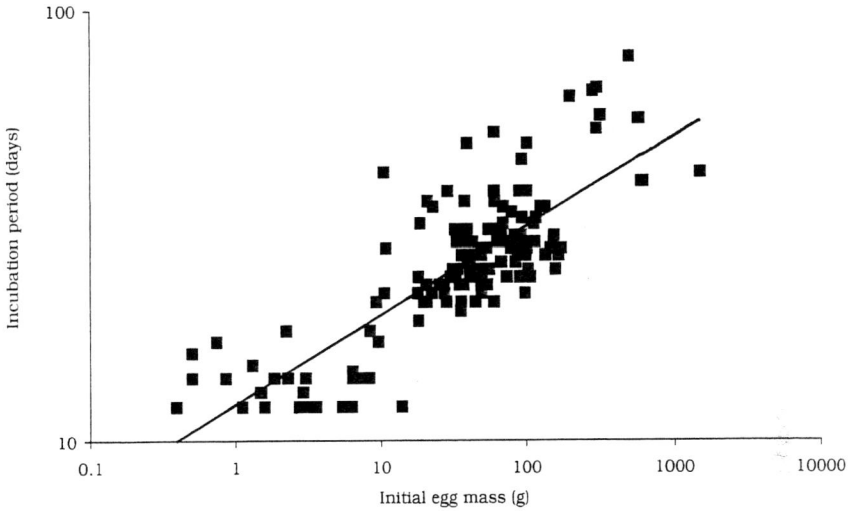

Summary

- Fertilisation takes place at the top of the oviduct within 15 minutes of ovulation.

- Embryonic development occurs in three phases: in the oviduct, differentiation during the first half of incubation, and growth during the second half of incubation.

- Life in the egg is impossible without the extra-embryonic membranes.

- Yolk sac retraction involves pulling in residual yolk into the body cavity in the days before hatching.

- Chicks rotate within the egg during hatching to break off the top of the shell.

- Incubation periods are variable within a species but between species correlate with egg size

5 - Behaviour of Adult Birds during Incubation – Creation of the Incubation Environment

The incubation environment in birds depends on the intimate contact between the bird's body and the egg(s). Indeed it is this contact that characterises incubation in birds although it is the interplay between the bird, the eggs and the nest that provides the ideal environment for embryonic development. This bird-nest incubation unit is critical to bird reproduction and each species has evolved its own specific sets of conditions matched to its local nesting environment. However, research has shown that incubation parameters in birds are incredibly conservative. The basic needs of the embryo in terms of maintenance of temperature, regulating humidity and gas exchange with the nest, and turning, have produced a set of conditions that are remarkably similar between species of vastly differing sizes and nesting environments.

This chapter describes firstly the behaviour of birds on sitting on eggs during incubation, secondly the factors affecting bird behaviour and finally how this activity brings about the characteristics of the nest environment, *i.e.* the correct temperature, humidity, gaseous environment and egg turning.

Initiation and maintenance of incubation behaviour

When incubation behaviour starts falls into two broad types depending on the relative size of the clutch. In the first type the birds wait until the clutch is complete before incubation starts and so embryonic development starts at the same time in all eggs and hatching takes place at the same time (*i.e.* is synchronous). This pattern is common in those species with precocial offspring and helps chick survival once the birds leave the nest. However, there is some limited evidence to suggest that during laying some incubation behaviour occurs and that embryonic development is asynchronous. Only during the last few days do individual embryos slow down or speed up their development in order to synchronise hatching (see Box 4.4, p. 62). More research is needed in this field to examine whether developmental asynchrony exists in nature.

The second group of birds initiate incubation with the penultimate egg of the clutch or earlier. Embryos in eggs laid early in the clutch get a head start and hatching is asynchronous. This pattern is more typical of species with semi-altricial or altricial young which do not leave the nest until fledging.

The factors that stimulate building of a nest and initiation of incubation behaviour have been the subject of considerable research and our understanding of the process is pretty good for at least a few species. "Broodiness" represents a move away from courtship and mating and is a response to various physiological and environmental cues. The presence of a nest or eggs may be important in some species. Similarly, the presence (or absence) of a mate can be important. Incubation behaviour is more typically initiated and maintained by hormonal activity in the bird.

Prolactin, a hormone produced in the brain, is the primary factor that induces broodiness. Often acting in conjunction with changes in other sex hormones, a rise in prolactin levels in the blood is the key to changes in behaviour that lead to incubation. The changes in the skin necessary for the production of a brood patch (see pp. 78–80) are also under the control of prolactin. High levels of prolactin in the blood are critical in maintaining incubation and chick rearing behaviours and are often maintained by feedback mechanisms associated with tactile and visual contact with eggs and chicks. In species that exhibit female-only incubation but shared rearing it seems that contact with eggs is not always necessary to maintain high levels of prolactin in the male birds. The mechanism by which prolactin levels are kept high in the non-incubating birds is not clear.

In many species predation of a clutch of eggs usually leads to a rapid drop in prolactin levels leading to a loss of broodiness and the nest is abandoned. This means that, should there be suitable conditions, the birds can re-start courtship and lay more eggs rather than being tied to unproductive incubation behaviour.

Basic behaviour of the adults during incubation

A pair of adult birds has a variety of options when it comes to incubating their eggs. Both or either gender can be involved in the incubation duties and birds practice every option. Indeed attempts to classify incubation behaviour on, for instance the basis of when and for how long each gender incubates, has highlighted the considerable degree of variation. Whether such detail is any practical interest could be questioned and so here I feel that it is best to stick with general trends.

Almost 50% of bird families exhibit "shared" incubation, where both the male and female spend time incubating the eggs. The proportion of time spent on the nest by the male and female is not necessarily equal in all cases but the male does take a significant role in keeping the eggs warm. Shared incubation is found in a wide range of bird types from penguins through to woodpeckers and songbirds.

In around 6% of bird families it is the male alone that incubates the eggs with no contribution from the female. This is typical in ratite birds (emus, rheas, tinamous and kiwis) with the exception of the ostrich in which incubation is shared between the male sitting at night and the female during the day. Male-only incubation behaviour is also observed in some shorebirds (notably the phalaropes) and in some families of the order Gruiformes. In all known examples of male-only incubation the females take no interest in either incubation or chick rearing.

Almost 38% of bird families exhibit female-only incubation with the male playing no role in incubation duties. This may be due to a complete lack of involvement in incubation or chick rearing duties (*e.g.* species like grouse that display at a lek [see pp. 6–7]) or the male has other duties during the incubation period (*e.g.* feeding the female, such as in some birds of prey). In other species the male has no role to play in incubation but is critical for successful rearing of chicks (*e.g.* many songbirds). In contrast to male-only incubation, behaviour patterns of the male and female birds during female-only incubation are rather flexible. Various factors will affect the behaviour of the male during incubation. For instance, high attentiveness in female birds of prey is usually only possible if the male provides food. In the meadow pipit the time taken by the female sitting on the eggs depends on the amount of feeding by her mate. If the male is inattentive or absent then the female has to forage for herself more and nest attentiveness drops.

For instance, in the European blackbird, the female takes charge of incubation sitting on the nest around 80–90% of the time. However, if the weather is cold the male will sit on the nest when the female is away foraging and the frequency of male sitting increases as the ambient temperature drops. The female appears to need the male to maintain egg temperature but only when the air is cold. In a similar vein, normally, the male kiwi incubates alone but in a species living in mountainous areas the cold climate means that the male has to feed more to survive. To prevent the egg from chilling the female is recruited to sit on the eggs during part of these periods of extended foraging.

Other incubation patterns exist but they are relatively rare. The red-legged partridge is reputed to lay two clutches and the female and male each incubate a clutch simultaneously. Just over 1% of bird families are brood parasites and neither gender plays any role during incubation (See Box 5.1, p. 69). In the majority of species it is the parents that incubate the eggs but incubation in a few species, which live in extended family groups, is shared between the parents and other related individuals in the group. There are still a few families of bird where we have not fully documented their reproductive biology (which accounts for the other 5% of bird families).

BOX 5.1 – BROOD PARASITES AND EGG-DUMPING

There are numerous species that forego the obligation of sitting on their own eggs. They prefer to lay their eggs in other birds' nests and leave these birds to incubate and rear their offspring. The cuckoos and cowbirds are the best known but there are types of finches and even one species of duck that are 'brood parasites'.

This behaviour has led to various adaptations of the parasitic birds that assist in them establishing their egg within a nest. It would be pointless to produce the egg if the host-bird were to instantly recognise the intruding egg and throw it out or abandon the nest. Such strategies include egg mimicry, changes in egg size and shell strength, timing of laying and relatively rapid embryonic development. The parasites do not always get their own way and the interaction between parasite and host is constantly evolving.

Cuckoos *etc.* have completely given up on incubation and chick rearing but another form of parasitism is widespread in birds. "Egg-dumping" involves a female laying an egg in another bird's nest even though she may have her own nest. This archetypal strategy of "not laying all of eggs in one basket" is a safeguard against loss the female's nest – at least there is a possibility that some of her offspring will survive in other nests.

Egg-dumping is normal practise in ostriches where females will establish nests but lay an egg in one or several other nests in the locality. Ironically, the female that sits on the eggs appears to be able to recognise her own eggs, although it is far from clear how she achieves this (see Box 3.2, p. 31) and exclude the eggs from other females from the nest. These lie in a ring around the nest and are invariably taken first by any predators that find the nest. Even if other female's eggs are incubated and hatch, these chicks will "dilute" the original female's offspring in the crèche and reduce the risk of predation.

Activity patterns during the incubation period

So what do birds do whilst incubating? The simple answer is: not a lot! Successful incubation means that the bird has to maintain contact with the eggs to transfer heat to (and from) the egg and so all it can realistically do is sit there. Indeed studies have shown that most birds will just sit there, often with their eyes closed, for very long periods of time. In many cases the birds will get up and leave the eggs usually to forage and other normal activities (an "incubation recess") before returning to the nest to start another period of incubation (an "incubation session"). The degree of this "nest attentiveness" is very important in the pattern of incubation in different types of birds (see pp.

71–78). At other times the bird will rise from the nest and turn the eggs (see pp. 94–99) or simply stand up, turn and re-settle on the eggs for no apparent reason (but see Box 4.4, p. 62). In waterfowl and gulls the process of sitting on the eggs after a recess is often accompanied by ritualised behaviours including tail wagging and calling. Otherwise whilst sitting the birds will preen themselves, tend the nest material or carry out other maintenance behaviours, such as regulating their own body temperature.

Indeed it is often the case that the prime consideration of the incubating bird is its own comfort. The location of the nest site will have taken into account of the need for the bird to sit motionless for many days. Nests hidden from sight will help reduce predation but will also help protect the bird from the elements. A good example of this is the Palestine sunbird. This tiny bird suspends its nests under buildings and other structures in locations that minimise the time that it will catch the full glare of the sun. Similarly, some hummingbirds locate nests under overhanging branches or rocks in order to minimise the exposure to the cold night sky. In more exposed situations the incubating birds are forced to endure the prevailing climate. Emperor penguins nest on the Antarctic ice sheet in winter and huddle together to minimise the effects of the biting wind (the egg sits on the male's feet covered by a pouch of skin). At the other extreme desert nesting species sit facing away from the sun during the heat of the day.

Many adult behaviours are geared to ensuring that incubation conditions are maintained for the eggs. Desert conditions pose the greatest problem due to high temperatures and dry air. At temperatures above normal incubation temperature (i.e. ~38°C) the bird sits tight on the eggs and instead of transferring warmth to the egg it actually removes the heat absorbed from the air by the egg. This is necessary because the egg has no means of regulating its own temperature and it relies on the adult to provide or remove heat. Some desert species will deliberately cool the eggs by dropping water on them or by sitting on them with wet feathers.

Incubation behaviour also includes defence of the nest and eggs. This can be by concealment or by adopting a nest site, which is difficult to reach (e.g. on a cliff face or top of a tree). Approach of a predator may force the bird to sit tight in the hope of not being detected, or leave quietly so as not to reveal the nest. Another strategy is to distract the predator away from the nest. Some birds, such as game birds, will fly away loudly calling to distract the predator. Other ground-nesting birds will leave the nest and then feign injury to attract the predator away before escaping itself once the danger of nest exposure is passed. Other species will actively defend the nest site by threats or by physically attacking a predator. Terns and skuas are notorious for their attacks on people near their nests. The female eider duck takes the rather extreme course of defecating on her eggs as she leaves the nest in the hope that the noxious smell will deter the predator.

Behaviour at the nest is often quite frenetic in colony nesting species of seabird that are forced to nest in close proximity because of lack of suitable nesting sites. Social interaction is necessary to defend the nest site and the nest itself before and during incubation. The crowded conditions usually mean that the paired birds have elaborate greeting displays in order to recognise each other.

Whilst behaviour during an incubation session is rather limited, during a recess the bird is free to do anything it wants. In those species incubating alone and with short recesses, then foraging is as a priority but in other species where incubation is shared the birds can perform other behaviours. Often these involve watching for predators or other threats to the nest.

Birds that take no role in incubation are essentially free to carry on their normal behaviours. In many species this often involves the male establishing new pairs and mating with more females. However, many male partners will pay a lot of attention to the nest and supplying the incubating female with food is an obvious benefit.

Nest attentiveness

Rarely do birds sit on their eggs all of the time during the incubation period. It is assumed that the eggs are always covered at night (100% attentiveness) but during the day the birds have periods (incubation recesses) when the eggs are left exposed or the bird is replaced by its partner. The proportion of time the bird spends on the eggs (an incubation session) varies between species and depends on upon a variety of factors. Before discussing these factors it is interesting to compare the incubation practises of different types of bird.

Most types of bird have average values for attentiveness at the nest and sitting on the clutch of over 90% of the daylight hours (Figure 5.1). However, for the swifts and songbirds attentiveness is only 70–75% of the daylight hours (Figure 5.1). Up to 30% of the time the eggs are left exposed in the nest, although the nest itself will almost certainly be hidden. This strongly implies that egg temperature does not remain constant throughout incubation and when egg temperatures are recorded then this is clearly the case (Figure 5.2). Egg temperature can drop considerably during these incubation recesses yet it seems to have little impact on the survival of the embryos. Presumably they are adapted to survive these conditions.

So what are the influences on how much time the parents sit on their eggs? The factors fall into two categories: 1) those that influence the broad pattern of attentiveness over the incubation period; and 2) those factors that influence

Figure 5.1. Average total nest attentiveness (by one or both parents) for different groups of birds. Values are percentages of the daylight hours spent on the eggs. Attentiveness during the night is assumed to be 100%.

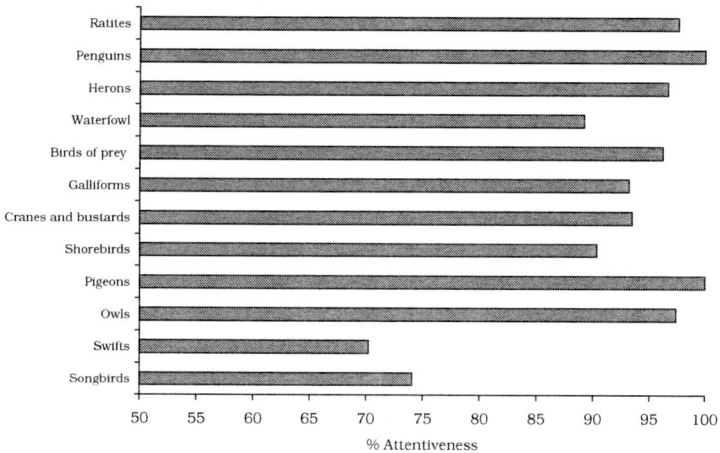

Figure 5.2. Recordings of egg temperature of the yellow-eyed junco over a period of 24 hours. Decreases in temperature indicate incubation recesses when the female left the nest. Data courtesy of Wes Weathers.

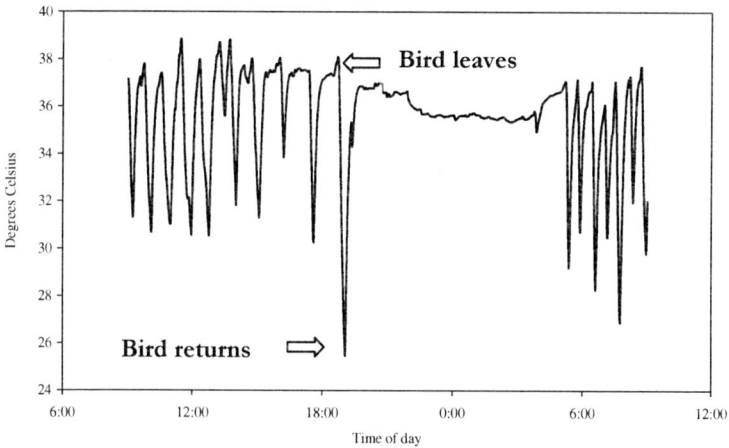

the attentiveness on an hourly or daily basis.

Firstly, it depends on who is incubating. Male ratites incubating alone generally have high %attentiveness although smaller shorebirds, in which the male incubates alone, tend to have lower %attentiveness (Figure 5.3). By contrast, female-only incubation exhibits a tremendous spread of values for %attentiveness (from <50–100%) with an average around 75%. However, there are many more species at the upper end of the range than there are at the lower end (Figure 5.3). Birds that share incubation have a much more restricted range for %attentiveness and the bulk of them are attentive for over 90% of the time (Figure 5.3).

The developmental maturity of the hatchlings is not particularly important in determining the spread of %attentiveness in those species with shared incubation. The lower values (< 80%) tend to produce altricial hatchlings but most species (altricial and precocial) have high attentiveness rates (over 85%). By contrast, in female-only species producing altricial young (typically smaller birds like songbirds) there is a wide spread of values (50–100%). By contrast, those species producing precocial young, which include species of waterfowl and grouse, then the range is more limited (75–100%) and most of the species have attentiveness values of over 80%.

For many species of small bird left alone to incubate the eggs alone there is a real conflict between the needs of the eggs to be kept warm and the needs of the bird to survive. Small birds have small fat reserves and so foraging is a critical aspect of their daily routine even when incubating. The result is "intermittent" incubation where the bird regularly leaves the nest for typically short periods (Figure 5.2). Only when there is provision of food by the male can the female spend more time on the nest. The need for foraging and other normal activities by the females mean that there is often a daily pattern to atte-

Figure 5.3. Distribution of the %attentiveness between different incubation strategies exhibited by birds. Re-drawn from data illustrated in *Avian Incubation: Behaviour, Environment and Evolution*, Oxford University Press 2002.

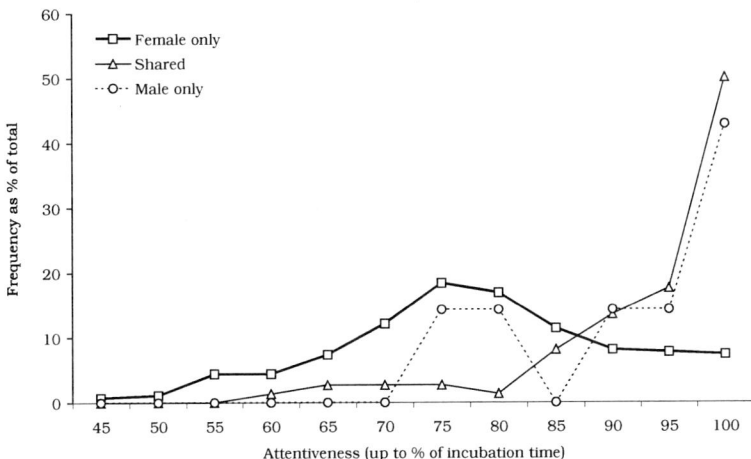

ntiveness with low attentiveness at the start and end of the day reflecting the need to break the nightly fast, and to stock up on food before the night, respectively.

However, as is explored in detail below, this not as simple as it first seems. Waterfowl and grouse also exhibit female-only incubation without male feeding yet they take only a few recesses that are usually quite prolonged (30–60 minutes). Hence, at 85% or more, attentiveness is much higher then in smaller birds. In birds of prey female-only attentiveness is often over 95% and can only be achieved by males providing food for the incubating females.

Nest attentiveness can also be affected by the stage of embryonic development. In some waterfowl and the Houbara bustard, %attentiveness decreases as the incubation period proceeds. By contrast, in smaller birds %attentiveness over the duration of the incubation period does not change.

On a daily basis there are other factors which affect nest attentiveness. Most of this research has been carried out in those species that have relatively low nest attentiveness. This is not surprising given that many species will be on the nest for well over 95% of the time, particularly if incubation is shared and so there is little scope for variation.

The most obvious condition affecting incubation behaviour is the prevailing climate. In particular, ambient temperature has important effects on the behaviour of the bird. Nest attentiveness for the Dead Sea sparrow drops as ambient temperature rises reaching a low of to less than 50% of the time

Figure 5.4. Correlations between maximum daytime temperature and the %attentiveness (black circles) and recess length (white circles) in the European blackbird. Lines are fitted by eye to show the trends.

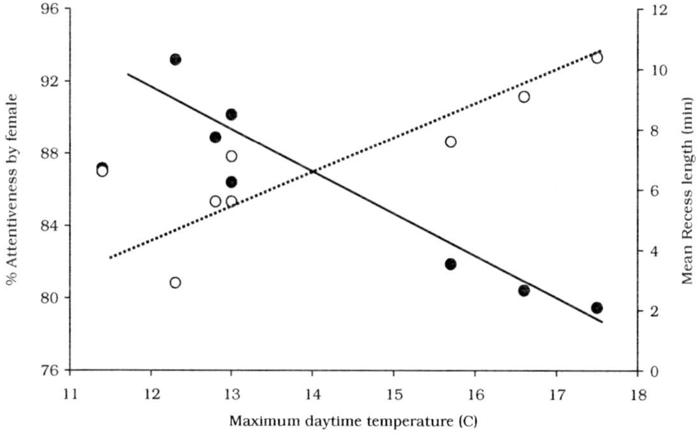

Figure 5.5. Average length of the incubation recess at different times of the day in the Houbara bustard. Photograph by DCD. Graph reproduced with kind permission of Science Reviews.

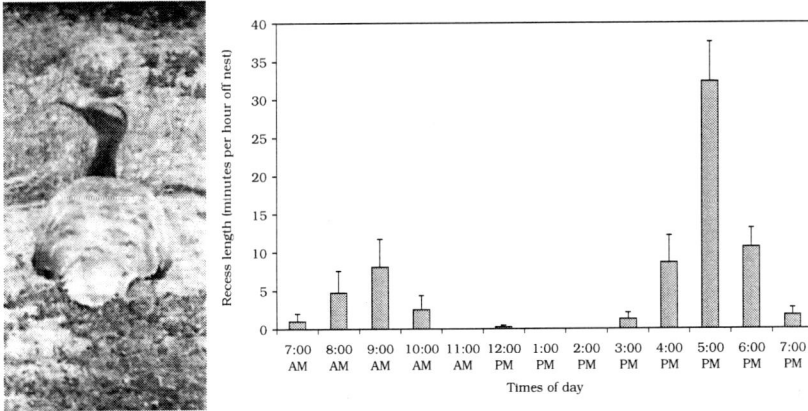

when the air temperature is 35°C. As the temperature continues to increase then nest attentiveness increases rapidly. The bird is responding to the need of the embryo to be protected from temperatures that are above its normal developmental range. The European blackbird also illustrates the relationship between temperature and attentiveness. Here as the average daily temperature increases the attentiveness drops because of an increase in the length of the incubation recesses (Figure 5.4). The female Houbara bustard takes a relatively short incubation recess (5–10 minutes) during the early morning when the air is cold. During the late afternoon, when the air is warmer it takes a much longer recess of 35 minutes (Figure 5.5). During the heat of day the female rarely gets off the eggs.

Nest attentiveness can be modified by a variety of factors as has been shown for other characteristic parameters of bird incubation (for instance water vapour conductance, weight loss and incubation period; see Chapter 3). Is there a single factor that is driving incubation behaviour in all types of birds? Recent research has suggested that the mass of the egg is an important influence on bird behaviour.

Egg mass and attentiveness

Research into the behaviour of birds during incubation has generated lots of data on the %attentiveness of a wide variety of species. Luckily it has been possible to match egg mass to most of these species and a simple plot of %attentiveness against egg mass has revealed a very interesting pattern (Figure 5.6). Small eggs of less than 1 g laid by, for example hummingbirds, are generally incubated for between 60–80% of the time but for eggs ranging from 1–10 g attentiveness ranges from less than 50% up to 100% (Figure 5.6). By contrast, egg masses from 10 to 100 g were incubated between 80–100% of the time. Above 100 g the range of attentiveness was even more restricted at between 90–100%. The ostrich, with the only egg over 1 kg, covers its eggs 100% of the time (Figure 5.6). This result suggests that as egg mass increases then the birds have to sit on the eggs more.

Clearly, the mass of a single egg is important if it is the only one in the clutch but many birds lay numerous eggs in the clutch. Once total clutch mass is taken into account then a similar pattern of attentiveness and egg mass is observed. Between 5–40 g of eggs in the total clutch then attentiveness varies from 50–100%. Above 40 g and up to a clutch mass of 1 kg, the attentiveness is between 80–100%. For clutches weighing over 1 kg (*i.e.* ratites) nest attentiveness is at least 95% of the time. The few eggs weighing between 10–100 g but from birds with low attentiveness (Figure 5.6) are actually from species with a single-egg clutch and so become typical once clutch mass is taken into account.

On average attentiveness in species with shared incubation is high (over 90%) although as egg mass increases then the birds become more and more attentive (Figure 5.6). For female-only incubators average attentiveness is

Figure 5.6. Relationship between initial egg mass (note Log plot) and %attentiveness in birds with shared incubation (black circles), male-only incubation (diamonds) and female-only incubation (white circles). Figure adapted from data illustrated in *Avian Incubation: Behaviour, Environment and Evolution*, Oxford University Press 2002.

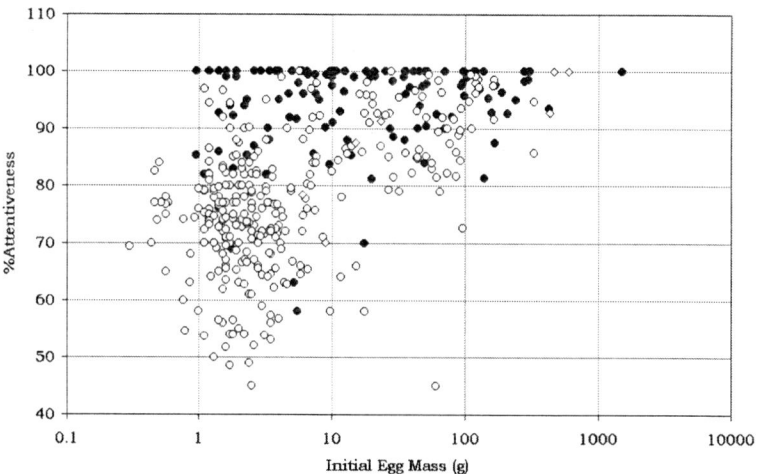

Table 5.1. Average % attentiveness of species exhibiting female-only or shared incubation.

Egg mass (g)	Female-only incubation (%)	Shared incubation (%)
1	72.0	92.7
5	75.9	96.0
10	70.3	95.9
50	89.5	94.3
100	85.1	97.1
500	No species	93.6

~75% for eggs in the range of 1–8 g but increase to ~85% for eggs in the range of 9–100 g. Above 100 g attentiveness is over 95%. Therefore, comparison of species exhibiting female-only or shared incubation shows clear differences in %attentiveness depending on egg mass (Table 5.1).

The idea that egg mass is correlated with attentiveness is supported by data for male-only incubation. The large eggs of ratites (450–600 g) are incubated for 99% of the time but as egg mass decreases then attentiveness goes down. For instance, the 23 g eggs of the Boucardi tinamou are incubated for 91% of the time whereas phalaropes, which lay eggs of only 6–9 g, have attentiveness values of only 70–78% (Figure 5.6).

Egg mass helps explain the difference between the attentiveness patterns of female-only incubators from different bird groups. Small songbirds, laying small altricial eggs have a wide range of attentiveness and the average value is ~75%. In the much larger eggs of waterfowl and grouse, producing precocial young, the range of attentiveness is more restricted and averages at around 90%.

The standard argument to explain the low attentiveness of species like songbirds is that the females have to forage during incubation and that this reduces the time that they can spend on the eggs. This certainly applies to the very small hummingbirds that are reliant on nectar to survive. For larger songbirds, factors like nest location or insulation and male-feeding allow the female to spend more time of the eggs. By contrast, bigger single gender incubators, like waterfowl, have more fat reserves and can spend more time on the nest. Unfortunately, there has been little attempt to explain why female waterfowl have few, long recesses compared with the numerous short recesses exhibited by female songbirds.

However, I believe that big birds have to sit on their big eggs because the thermal characteristics of the eggs "demand" it. The cooling rates of the bigger eggs mean that the birds have to sit on them for longer periods in order to ensure that the temperature is correct. If they get off them then it takes a long time to warm them back up. By contrast, small eggs (<10 g) have different thermal characteristics that cause them to cool rapidly but they also warm up quickly. Thus small eggs allow the birds to fly off on a regular basis. This idea is described further in a later section of this chapter (see pp. 80–87).

In conclusion, as for other incubation parameters in birds (see pp. 34–35) it would seem that incubation attentiveness is strongly correlated with egg size. How the birds behave to provide heat to maintain the appropriate egg temperature is maintained by the birds depends largely on the thermal characteristics of the eggs.

Maintenance of egg temperature

Bird incubation is characterised by the fact that the provision of heat required to support embryonic development is supplied by an adult bird resting a part of its body on top of the eggs. Typically, the bird sits with its abdomen on top of the eggs and there is often, but not always, a specialised area of skin (the brood patch), which improves heat exchange. In some species other body parts are used for heat transfer. In pelicans and gannets, incubation involves the adult wrapping its webbed feet around the egg. In the Emperor and King penguins, the extremely cold nesting conditions mean that the birds have to incubate the eggs resting on top of the feet and covered by a flap of skin. The megapodes of Australasia and SE Asia are exceptional birds because they have evolved an incubation system where the eggs are buried in volcanically heated sand or in mounds of rotting vegetation and the physical contact with the adult is lost (see Box 5.2, p.79).

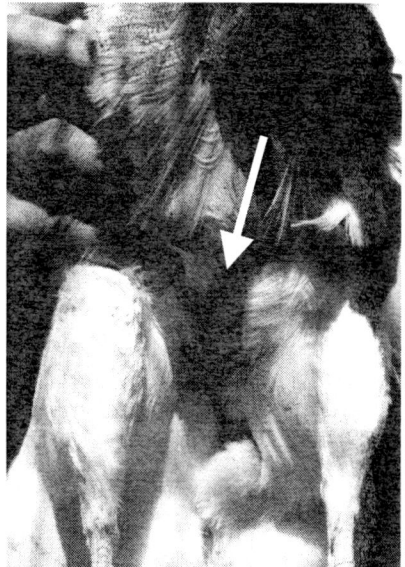

Figure 5.7. The underside of a female Houbara bustard showing the bare skin of the brood patch (see arrow). Photograph by DCD.

The brood patch

A brood patch (Figure 5.7) is not a necessity for efficient incubation but it is present in a wide range of bird species. The absence of a brood patch certainly does not indicate that the bird does not incubate. For instance, in many species, such as the zebra finch, the male has significant incubation duties but only the female develops a brood patch. Neither male nor female ostriches develop brood patches yet they share incubation duties.

Located on the lower surface of the abdomen (Figure 5.8) the brood patch is a localised area of skin characterised by loss of feathers, oedemic, thickened skin, and a rich supply of blood vessels. All of these alterations in the skin are brought about by hormonal changes associated with egg production and laying. In some species the presence of the eggs in the nest stimulates formation of the brood patch. Its persistence during incubation is often due to the physical contact with eggs. Many birds have a large central brood patch that

---

BOX. 5.2 – INCUBATION IN THE MEGAPODES

Megapode birds are unusual in that they have dispensed with contact incubation and have sought other sources of energy to keep their eggs warm during incubation. The eggs are either buried in mounds of rotting vegetation built, and tended, by the adults or in burrows dug into sand that heated by subterranean volcanic activity. In mounds the correct temperature for embryonic development is caused by heat produced by microbial fermentation of the vegetation but the adult has to tend the mound, adding and removing material, in order to maintain the correct temperature. Burrow-nesters tend to abandon the eggs assuming that the nest temperature will remain constant. This behaviour does not represent an ancestral condition for incubation in birds but rather it is a regression to the incubation habits of the reptiles.

The habit of burying eggs has meant that the eggs and embryos have had to adapt to the new environment. The eggs are large relative the body size of the adults and are exceptionally rich in yolk. In mounds this is characterised by very high humidity, low levels of oxygen and high levels of carbon dioxide. The eggs have relatively thin shells that are initially low in porosity but this increases considerably as incubation proceeds because the embryo etches away the inside of the shell as it removes the calcium for its bones. Water loss from the eggs is relatively low (not surprising given the high humidity) but increases during the second half of incubation as the shell is thinned. The eggs are also not turned at all during incubation and the length of incubation is long compared to eggs of a similar size from other non-megapode species.

Megapode hatchlings are unusual in that they are fully feathered and totally independent at hatching. After they have dug their way out of the substrate surrounding the eggs, they are able to fend for themselves and can fly within a day!

Figure 5.8. Diagrammatic representations of the area and relative size of the brood patch (black area) of Californian quail (a), red-necked grebe (b) and white-crowned sparrow (c). Stippled areas indicate feather tracts. Drawings reproduced with permission from *Avian Incubation: Behaviour, Environment and Evolution*, Oxford University Press 2002.

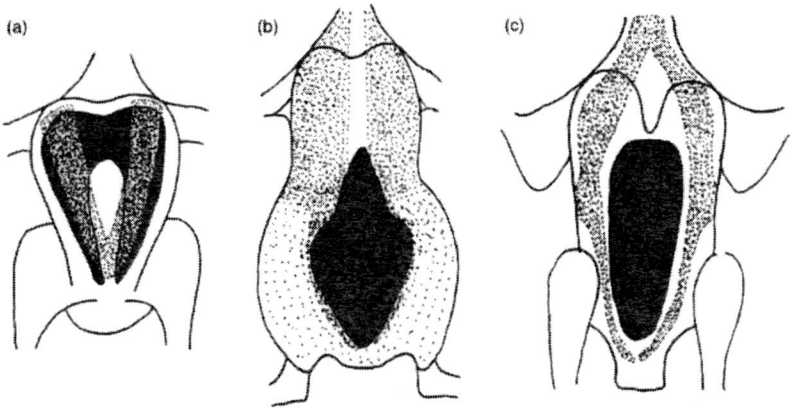

can accommodate several eggs (Figure 5.8) but in other species (*e.g.* gulls) each egg has its own brood patch on the abdomen, which makes clutch size quite critical for efficient incubation. Although not measured in a lot of species it would seem that the brood patch covers 15–20% of the eggshell surface.

Contact with the brood patch is the means by which heat is transferred from the body of the bird to the eggshell and hence into the egg contents by conduction. This function (although crucial it does have downsides [see Box 5.3, p. 81]) has meant that control of the temperature of the egg is critical and the bird is able to sense when egg temperature is not correct. Experiments have shown that cooling of an artificial egg (see Box 5.7, p. 98) causes an increase in blood flow through the brood patch – thereby bringing more heat from the body core. Prolonged cooling of an egg causes the bird to increase its oxygen consumption and heat production. When the artificial egg is infused with warm water then the brood patch absorbs the heat and the sitting bird begins to pant in order to lose the additional heat energy. Cutting the nerve supply of the brood patch can also adversely alter the behaviour of the incubating bird.

Heat transfer to the egg

As will be shown in Chapter 7 an egg in an incubator is incubated by convection (transfer of heat via the air) and in force-draught machines the egg has a uniform temperature profile, *i.e.* the top, centre and bottom of the egg are the same temperature. In a nest, the egg is heated by conduction of heat from

---

BOX 5.3 – CONFLICTS BETWEEN CONTACT INCUBATION USING A BROOD PATCH AND EMBRYONIC DEVELOPMENT.

Pressing a brood patch to the top of an egg is an efficient method of heat transfer to the embryo but it can cause problems. There is ~20% coverage of the eggshell by the brood patch and this means that the surface area for gas exchange is reduced. The skin effectively blocks the pores in the shell and prevents oxygen diffusing across to the embryo. Although this may be important early in incubation because it prevents oxygen toxicity it can cause localised hypoxia and hypercapnia later in incubation. Recent work has shown that prolonged contact with a brood patch also diminishes the density of blood vessels in the chorio-allantoic membrane – the longer the contact the more the blood vessels regress. The brood patch also effectively cuts down the area for diffusion of water vapour out of the egg.

It seems, therefore, that constant contact with the brood patch would not be in the embryo's interest. Recent research has shown that there is a mechanism for the embryo to communicate with the adult. Nitric oxide (NO) is produced by eggs and can be detected by the brood patch skin. Experiments have shown that factors like egg temperature and hypoxia elicit changes in NO production which may modify the attentiveness or turning behaviour of the adult.

---

the brood patch and is characterised by a temperature gradient from the warm top of the egg to the cooler bottom of the egg. In most still-air incubators a temperature gradient is also generated within the egg (see p.115).

This difference in the way heat is applied to the egg needs careful consideration when describing how embryo temperature is maintained. This concept is important because it is the temperature of the embryo, and not egg temperature, that is critical for its pattern of development. Records of egg temperature show an average of only 35.7°C compared with an average body temperature of 40.7°C in the same species. For a variety of reasons (see p. 25) the embryo is located at the top of the egg for the majority of the time. Hence, it is close to the shell in contact with the brood patch. Therefore, even though the temperature gradient in an egg during the early stages of incubation means that the centre of the egg is relatively cold, the embryo is going to be close to the brood patch temperature.

The point that average egg temperature is not important is illustrated very well by the temperatures recorded in ostrich eggs during natural incubation (Figure 5.9). At the start of incubation the embryo lies at the top of the egg within a few mm of the brood patch, which is itself close to the body temper-

Figure 5.9 A diagrammatic representation of the temperatures recorded at the brood patch and at various positions in an ostrich egg.

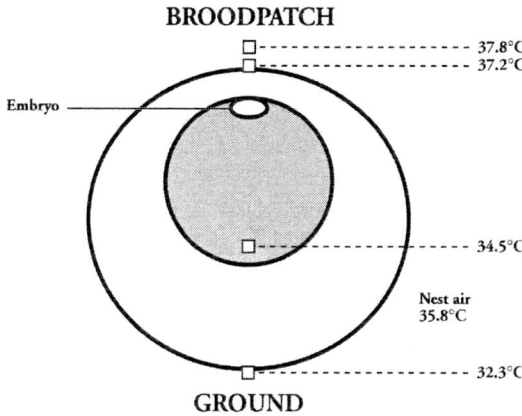

BROODPATCH

37.8°C
37.2°C

Embryo

34.5°C

Nest air
35.8°C

32.3°C

GROUND

ature of the bird (37.8°C). The centre of the egg is at 34.5°C and the bottom of the egg, adjacent to the nest is only 32.3°C. Nest air temperature is 35.8°C. Hence, embryo temperature is above 37.0°C despite the lower temperatures elsewhere in and around the egg.

The magnitude of the temperature gradient depends upon the size of the egg and the stage of incubation (Figure 5.10). At the start of incubation the heat transferred from the brood patch in, for example, a 2 g songbird egg does not have a great distance to travel. Therefore, the temperature gradient in the egg is relatively shallow – the bottom of the egg is similar to the temperature recorded at the top of the egg. During the incubation session the bird is transferring heat energy into most of the egg.

As egg size increases the temperature difference between the top and bottom of the egg (and other parts of the egg distant from that immediately next to the brood patch) increases. In a 50 g fowl egg the temperature gradient is more pronounced and energy transferred from the brood patch warms only the top part of the egg.

The temperature gradient is at its greatest in the 1,500 g ostrich egg with the area immediately next to the brood patch being the warmest with the core being cool (Figure 5.10). At the start of incubation energy transfer from the adult's skin will make little impact to the temperature of the ostrich egg as a whole (Figure 5.9).

The size of the temperature gradient will also depend on the amount of the shell surface covered by the brood patch and the nest temperature – the cooler the nest air then there will be more heat lost and the gradient will be greater. Insulation within the nest will also prevent heat loss and reduce the temperature gradient.

Embryonic development has been shown to influence egg temperature in two ways. The first is the embryo produces heat as a waste product of its metabolism and this heat output increases as the embryo gets bigger. Therefore, during the second half of incubation there is sufficient heat for the egg temperature to be maintained for a short period should the incubating bird leave the nest. The second factor affecting egg temperature is development of the blood circulation system in the extra-embryonic membranes. The blood distributes warmth from the area immediately below the brood patch around the egg where it can be lost through the shell.

The consequence of these events is the increasing core temperature of the egg and the loss of temperature gradients between the brood patch and the more distant parts of the egg. Again this is illustrated well by the ostrich egg. Core egg temperature is only 34.0–34.5°C by the end of the second week of incubation but as development proceeds the core temperature increases reach-

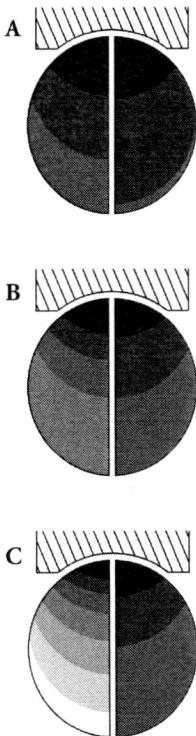

Figure 5.10. Diagrammatic representation of the temperature gradients within eggs of different sizes (seen in cross section) at the start (left semicircle) and end of incubation (right semicircle) under a brood patch (hatched area). A – 2 g songbird egg, B – 50 g fowl egg, C – 1,500 g ostrich egg. In each image the darkest grey represents the warmest part of the egg with decreasing darkness indicating increasingly lower temperatures. In the smallest egg (A) most of the egg is close to brood patch temperature at the start and end of incubation. In the 50 g egg (B) there is a pronounced temperature gradient in the egg at the start of incubation which is only partially lost by the end of incubation. In the largest egg (C), the temperature gradient at the start of incubation is very pronounced but at the end of incubation temperature distribution within the egg is more even. Therefore, heat from the brood patch is distributed throughout much of the small egg but very much less so in the large egg. The temperature gradients are largely destroyed by the embryonic circulation and metabolic heat production as the embryo develops.

Figure 5.11. Changes in the core temperature of fertile (white circles) and infertile (black squares) ostrich eggs during incubation under a brood patch.

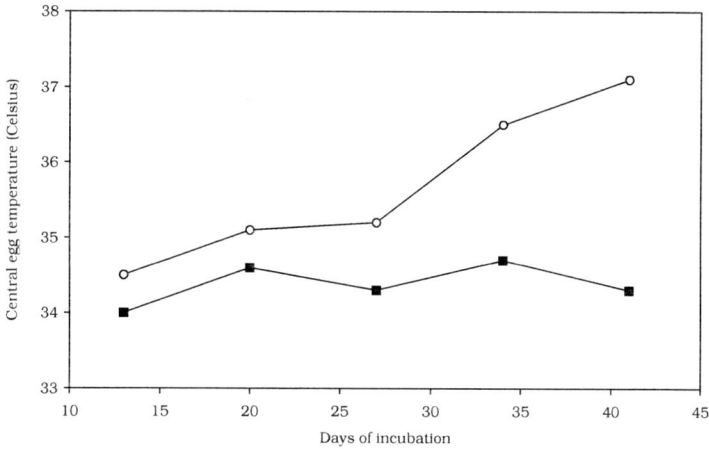

ing 37.1°C at hatching (Figure 5.11). The temperature gradient present in the egg at the start of incubation is all but abolished at hatching (Figure 5.10). By contrast, core temperature in infertile ostrich eggs does not increase because there is no embryonic growth, circulation or metabolism (Figure 5.11).

This pattern is repeated in smaller eggs with 50 g fowl eggs exhibiting a relatively uniform temperature profile (Figure 5.10) at the end of incubation. In the 2 g songbird egg the whole egg is almost at the brood patch temperature at hatching (Figure 5.10).

Egg temperature is, therefore, not simply a feature of brood patch contact but is influenced by ambient temperatures, egg size, the presence of the embryo, the degree of blood circulation and the metabolic heat output. So during the first third of incubation heat from the brood patch is not supplemented by any metabolic heat and is only lost to cooler parts of the egg through conduction through the yolk and albumen. During the middle third of incubation, embryonic metabolism is negligible but the extensive extra-embryonic circulation will conduct heat from the area immediately next to the brood patch to cooler part of the egg. During the last third of incubation heat from the brood patch is supplemented by energy from embryonic metabolism, all of which is being distributed around the egg by the blood supply.

Theoretical studies have shown that energy transfer between the brood patch and the egg is also affected by mass. Heat transfer into smaller eggs is more efficient and faster than for larger eggs. A 15 minute incubation session may warm up a 2 g egg completely but only warm the top of a 50 g egg.

So what happens when the bird gets off the nest? Contact with the skin

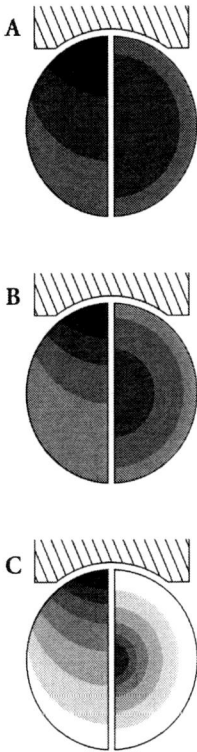

Figure 5.12. Diagrammatic representation of the temperature gradients within eggs of different sizes (seen in cross section) at the start of incubation with the bird on the egg (left semicircle) and off the egg for ~10 minutes (right semicircle). The brood patch is shown by the hatched area and the embryo is located at the top of the egg. In each image the darkest grey represents the warmest part of the egg with decreasing darkness indicating increasingly lower temperatures. In the smallest 2 g songbird egg (A) most of the egg is close to brood patch temperature when the bird is on and off the egg. In the 50 g fowl egg (B) when the bird is absent the centre of the egg is much warmer than its surface. In the largest 1,500 g ostrich egg (C), the temperature gradient at the start of incubation is very pronounced with the bird on the egg. Once the brood patch is removed the temperature gradient remains but shifts so that the egg core is warm but the surface is cold. Therefore, as egg mass increases heat from the brood patch is lost to both the surface and the core and the embryo, located at the top of the egg will cool significantly. These temperature gradients are largely destroyed by the development of embryonic circulation and by metabolic heat production.

and the egg's source of heat is lost and the upper surface is suddenly exposed to cool air. In a small egg (2 g) the surface begins to cool, assuming that ambient temperature is below the egg temperature, and the heat energy stored in the egg contents during the incubation session begins to leave the egg. There is not much of a temperature gradient from the core to the surface (Figure 5.12) and heat loss is mainly from the exposed upper surface. Energy flow from the core of the egg means that the temperature of the embryo is still being maintained by the energy previously invested in the egg by the sitting female.

In larger eggs removal of the brood patch leads not only to heat loss through upper shell but there is also movement of heat down into the egg. Significant temperature gradients form between the core and the surface (Figure 5.12) which persist for at least 50% of the developmental period until the embryo is large and well developed. When the bird leaves the heat energy in the warmest part of the egg, which is immediately below the brood patch, moves away by conduction to the centre of the egg and the embryo at the top of the egg is rapidly cooled as heat is also lost to the air. Only after a long while will the energy stored in the egg start to move back outwards towards the shell and begin to slow down the rate at which the embryo cools. The longer the bird sits on the egg the higher the investment of energy in the egg. As egg size increases further this phenomenon is exaggerated and is at its most extreme in the ostrich egg (Figure 5.12).

These thermal characteristics mean that efficient heat transfer is achieved after different session times for various sizes of eggs. Theoretical studies predict that incubation sessions for small eggs (1 g) will be short (10 minutes) but as egg size increases the session length increases to 200 minutes for a 100 g egg.

This observation helps to explain the differing %attentiveness of birds incubating eggs of various sizes. In effect birds incubating big eggs have to remain on them for longer periods to achieve efficient heating of the contents and long incubation recesses are possible because of the energy stored in the egg and its delayed release from the core. As eggs decrease in size, they become easier to warm up and so session and recess lengths can be shorter. Small birds go off their nests to forage more than bigger birds because the thermal characteristics of their small eggs allow them to do so. Bigger eggs need to be more constantly warmed to ensure embryo temperature is maintained.

The large eggs of ratites mean that the male-only incubators have to sit almost continuously to prevent chilling of the embryo. They are unable to forage and have to fast surviving on fat reserves established prior to the breeding season. By contrast, the ostrich is much more an opportunistic breeder, is unable to predict a particular season for breeding and cannot lay down fat reserves. As a result maintaining embryo temperature requires incubation to be shared by both parents.

Data on the lengths of incubation sessions and recesses of birds show that, as predicted from theoretical considerations, there is a strong relationship between egg size and attentiveness. For those species with female-only incub-

Table 5.2. Average values for incubation pattern for female-only incubators in four groups of birds. N = sample size; IEM = initial egg mass; CM = total clutch mass.

Bird group	N	IEM (g)	CM (g)	Session length (min)	Recess length (min)	Attentiveness (%)
Waterfowl	34	93.9	572.5	556.8	57.6	88.5
Grouse	5	40.5	369.1	910.1	49.1	93.5
Passerines	101	2.9	14.4	35.0	9.8	75.0
Hummingbirds	9	0.6	1.2	19.9	9.0	72.7

ation as egg mass (and clutch mass) increases there is a general trend for longer sessions and recesses and higher attentiveness (Table 5.2). However, the rate of increase in session length is twice that of the rate of increase for incubation recess length and so the increase in attentiveness is due to longer sessions rather than shorter recesses. This fits the prediction for the length of incubation sessions based on the thermal characteristics of the eggs.

In summary, the process by which the egg is heated depends upon contact between the shell and part of the bird's body, usually a brood patch. The actual temperature of any part of the egg depends on its position relative to the brood patch, egg size, ambient temperature, and stage of embryonic development. It takes a short period to warm through small eggs allowing the incubating bird to leave them on a regular basis without too many problems. As egg mass gets bigger then the thermal characteristics of the egg mean that the birds have to be more attentive and the behaviour pattern change. The biggest eggs have to be covered for 95–100% of the time and the incubation duties are either shared or the lone incubating bird fasts until the clutch is hatched.

Provision of nest humidity and ventilation

The bird-nest incubation unit creates a micro-environment for the eggs that can be influenced by the prevailing environmental conditions, the bird and the eggs. Maintaining the egg at a temperature higher (or lower) than the environment itself affects the gaseous environment around the eggs. For instance, warm air holds more water vapour than cold air for the same relative humidity. In this section the gaseous environment of the nest is considered

Figure 5.13. Number of species losing differing amounts of weight during incubation.

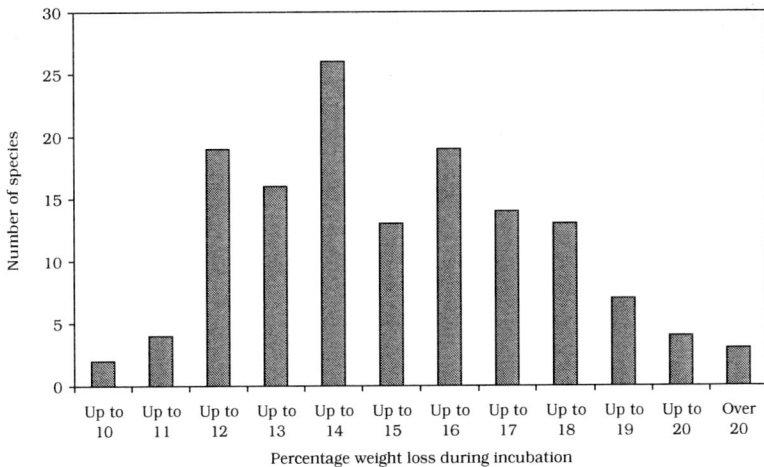

Percentage weight loss during incubation

and describes two main components: humidity, and the respiratory gases oxygen and carbon dioxide. The former is critical in regulating the rate of weight loss from the egg and so the importance of weight loss during incubation is discussed prior to a description of our understanding of humidity conditions in the nest. Thereafter, regulation of oxygen and carbon dioxide levels in the nest air is discussed.

Weight loss from bird eggs during incubation

Without exception bird eggs lose weight during incubation, including in the buried eggs of the megapodes (Box 5.2, p. 79). Whilst there is loss of carbon dioxide during development the mass of this gas is balanced by the amount of oxygen taken up into the egg by the embryo. Hence weight loss during incubation is equal to loss of water vapour by diffusion through eggshell pores and is controlled by the porosity of the eggshell and humidity outside the shell (see also pp. 33–36).

Studies into the rates of weight loss during natural incubation have shown that there is a wide variation in the average percentage of water lost for individual species (from 9–23% of initial egg mass) but most values fall between 10–20% weight loss (Figure 5.13). The average weight loss to external pipping of 140 species is 14.6% of initial egg mass (standard deviation = 2.6). On the

Table 5.3. Average values for initial egg mass (IEM, g), incubation period (Ip, days), daily water loss (M_{H_2O}, mg/day) and percentage weight loss (%WL) for different groups of birds.

	IEM	Ip	M_{H_2O}	%WL
Ostrich	1368.0	42.0	4300.0	13.2
Penguins	113.0	36.5	409.0	13.4
Grebes	60.9	30.5	265.5	14.4
Albatrosses	115.0	53.1	274.6	15.6
Pelicans	68.8	41.0	216.7	13.1
Herons	34.7	23.3	208.8	14.1
Waterfowl	105.0	25.4	605.4	14.7
Birds of prey	62.0	32.8	260.3	15.7
Game birds	41.7	23.3	281.7	15.3
Cranes, rails and bustards	38.6	21.7	203.4	13.3
Shorebirds	47.1	27.3	247.9	14.3
Owls	25.3	29.2	130.3	16.2
Nightjars	5.8	19.0	40.0	13.1
Woodpeckers	5.4	12.0	81.0	18.0
Songbirds	3.2	13.1	36.6	15.0

Figure 5.14. Relationship between daily weight loss and egg mass in 140 bird species.

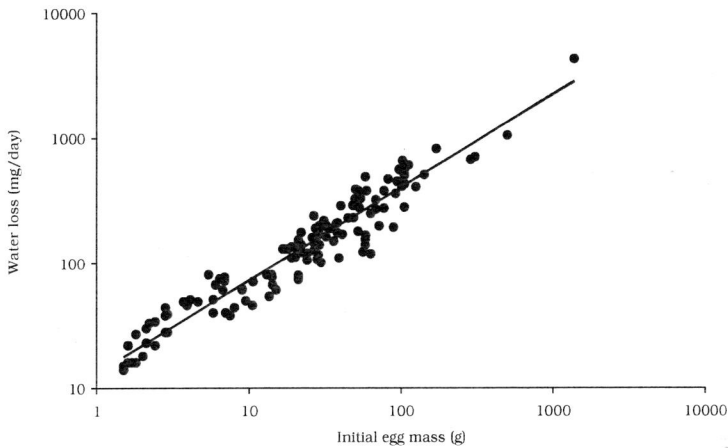

whole this average value is reflected in the range of different bird groups represented in the sample irrespective of egg size or incubation period (Table 5.3). It is important to note that these values are averages for numerous eggs and the actual weight loss of an individual egg is of less importance. Indeed individual eggs in wild clutches have been seen hatching after losing from 9 to 33% of their initial mess. In general terms a weight loss of between 10–20% will not cause an individual embryo any great problems (see Box 5.4, p. 90).

Indeed the extensive work of Herman Rahn, Amos Ar and Charles Paganelli and their co-workers during the 1970s–1980s showed quite clearly that the eggshell characteristics and other incubation parameters all scaled to egg mass (see pp. 33–36). Hence daily water loss scales with egg mass (Figure 5.14) and this relationship is linked to the needs of the embryo to be supplied with oxygen. Basically, bird eggs are very similar in their incubation requirements and factors like incubation temperature, nest humidity, weight loss and respiratory gas exchange are all inter-linked and related to egg mass. The importance of the correct level of weight loss is described in Box 5.4, p. 90).

Maintenance of nest humidity

So if weight loss is so critical how does the bird regulate humidity to ensure the correct environment? The relationship between nest humidity and ambient humidity has been studied in only a few species. The results of early studies were interpreted as suggesting that the incubating bird was actively regulating the humidity of the nest. However, more research has shown that

90

BOX 5.4 – THE IMPORTANCE OF CORRECT WEIGHT LOSS DURING INCUBATION

Birds exhibit a range in developmental maturity at hatching from fully feathered precocial species to naked altricial types. These extremes represent different developmental stages at hatching and the hatchlings have different percentage water contents – high in altricial and low in precocial young (see pp. 38). These differences are also seen in the water content of the yolk and albumen at egg laying. Embryonic metabolism creates water as a by-product, which adds to the water content of the egg contents. Therefore, one function of weight loss is to help balance the water content of the hatchling with that of the initial egg contents.

A second role is to allow for the formation of an air space within the egg. This causes an asymmetrical pattern in the density of the egg causing it to tip up (Figure 5.16) assisting in embryo orientation. More importantly, the air within the air space is used by the embryo to fill its lungs and air spaces during the hatching process. This allows these important organs to be fully functional when the embryo breaks the shell and begins the final stages of hatching.

Failure to lose the appropriate amount of water during incubation leads to problems with embryonic viability. Excessive weight loss leads to dehydration of the egg contents that can affect the embryo's metabolism and ultimately prevent it from developing normally. Insufficient weight loss is often a symptom of low shell porosity (see pp. 167–171) but can cause problems for the embryo as hatching approaches. Excess water has to be stored under the skin and in muscles and the bloated embryos are often unable to move in the egg, which prevents pipping and hatching. It has been shown in the ostrich, that the more weight the egg loses during incubation then the nearer to the equator the chick pips! Oedemic chicks are physically unable to pip anywhere but near the blunt pole of the egg.

---

over the incubation period nest humidity is simply elevated above that of the environment (Figure 5.15). The bird-nest incubation unit forms a microenvironment in which humidity, lost from the bird's skin and from the eggs, can build up to a level above that recorded in the environment. On the whole fluctuations in the prevailing humidity are reflected in the humidity under the bird (Figure 5.15).

This concept is reinforced by other data for average values of ambient and nest humidities. In general humidity under a bird is elevated above that of the surrounding fresh air (Figure 5.16). At high ambient humidities the nest hum-

Figure 5.15. Measurements of average humidity recorded in nests and the local environment in four species of bird. Vertical axes are values for humidity measured in Torr. Reproduced with permission from *Egg Incubation: Its Effect on Embryonic Development in Birds and Reptiles*, Cambridge University Press.

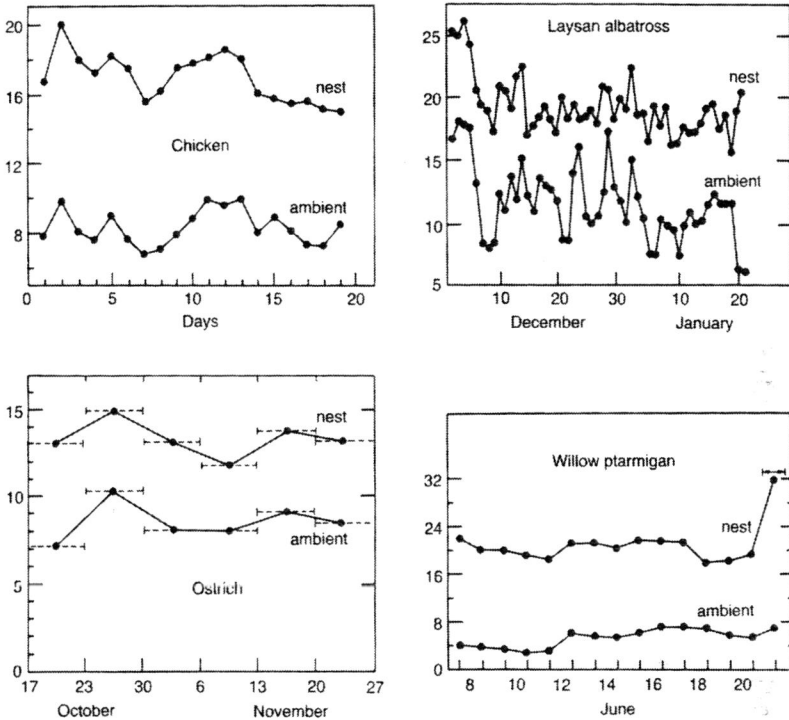

idity is only slightly elevated but as the ambient conditions get drier, nest humidity is maintained at relatively higher levels and in species studied to date rarely falls below 15 Torr (Figure 5.16).

Various studies have artificially altered the nest humidity of incubating birds in order to see whether the birds modified their behaviour to adjust egg weight loss. In domestic fowl and species of songbird, artificial regulation of humidity has no effect on bird attentiveness or nest ventilation behaviour and weight loss from the eggs is modified according to the changed humidity conditions.

Although the bird may not actively regulate the humidity in its nest, and although it raises humidity above ambient, it does play a key role in determining the humidity in which the eggs are incubated. Choosing the correct nest site is critical in achieving the optimum weight loss during incubation and

92

Figure 5.16. Plot of ambient water vapour pressure versus nest water vapour pressure for a variety of birds. The solid line indicates the trend represented by the points whereas the dashed line indicates equality between the two variables.

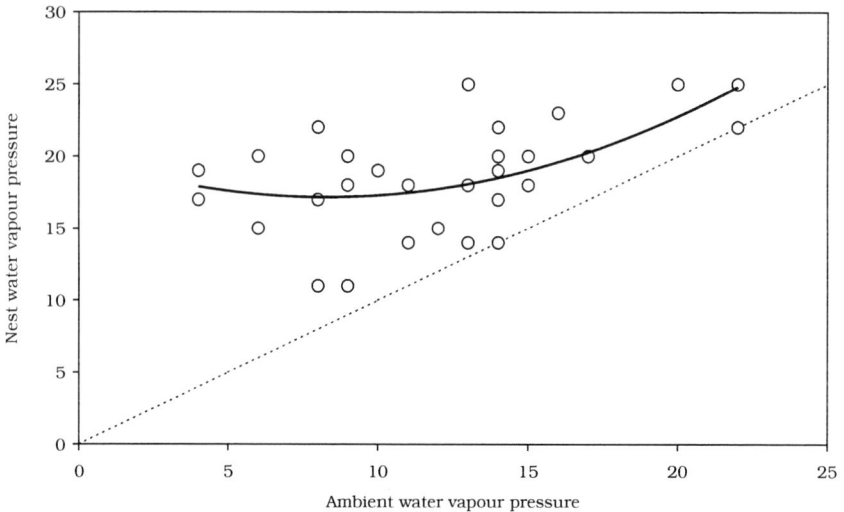

BOX 5.5 – INCUBATION AT ALTITUDE THE LINK BETWEEN LOCAL ENVIRONMENT AND EGGSHELL POROSITY

Birds nest in a variety of adverse climates but those living in mountainous areas face a particular challenge to successful breeding. Although faced with problems of cold temperatures the major problem is low barometric pressure, which increases the rate of diffusion of water vapour and respiratory gases. The main response is the reduction in eggshell conductance, by a reduction in pore numbers (rather than shell thickness), which counteracts the more rapid diffusions rates. The lower levels of oxygen at altitude mean that hypoxia is a particular problem of altitude embryos but most species studied appear to have acclimated to the conditions.

There is some evidence from domestic fowl that females are able to adapt their eggshell conductance in response to being transported to high altitude but work with other species has suggested that this is not a widespread ability. It is likely that the factors affecting eggshell thickness and pore number are genetically controlled and cannot be easily modified.

eggshell porosity is linked to a particular environment (See Box 5.5, p. 92).

Recent work on the bird-nest incubation unit has placed a lot of emphasis on the role of the nest in controlling humidity for the eggs. In simple terms the nest wall also allows loss of water vapour down a diffusion gradient from the more humid nest to the less humid atmosphere. Previous studies have shown that the eggshell appears to be the primary controlling factor in determining water loss from the egg to the atmosphere (via the nest). However, the fact that the brood patch covers perhaps up to 20% of the shell means that eggshell conductance and nest conductance are more equally balanced.

Therefore, the humidity environment provided by a bird for its eggs is a combination of behavioural factors (*e.g.* nest site selection, attentiveness patterns), physical factors (nest structure and conductance) and physiological factors (*e.g.* porosity during eggshell formation). These factors also affect oxygen and carbon dioxide, other gaseous components of the environment critical for embryonic survival.

Gaseous environment

Oxygen (O_2) is essential for life and so the eggshell has evolved pores which allow the exchange of gases during incubation (see pp. 32–37). Oxygen diffuses into the egg because the embryo uses the oxygen and lowers its partial pressure inside the egg creating a difference in partial pressure across the shell. Conversely, carbon dioxide (like water vapour) has a higher concentration inside the egg because it is a waste product of the embryo's metabolism. There are negligible amounts of carbon dioxide (CO_2) in the air and so it diffuses out of the egg. As development proceeds the embryo's requirements for oxygen increase rapidly and increasing CO_2 production. If the bird sat tight all of the time then the nest microclimate would change: the humidity and CO_2 levels would rise and the oxygen levels would fall leading to a stale environment for the embryo. The bird counteracts this situation by periodically rising from the nest exposing the eggs and effectively allowing the nest microclimate to be destroyed.

The rate at which birds get up from the nest varies. In albatrosses nest ventilation occurs every 46–78 minutes. In gulls the bird gets up around every 11 minutes whereas penguins get up every 20 minutes. Of course every time the bird turns its eggs (see pp. 97–98) will also expose the clutch to the prevailing atmosphere. Of course the location and structure of the nest will influence nest ventilation. If a nest has good thermal insulation then it will be less affected by wind than a nest constructed of an open stick mesh.

One problem in understanding the role of these gases in embryonic development is that very little work has been done to measure O_2 and CO_2 concentrations in bird nests. This is mainly due to the practical difficulties of taking the measurements. An old study (from the 1920s) of an incubating domestic fowl showed that CO_2 levels were elevated to ~1% (compared with 0.03% in fresh air) by the end of incubation. The decline in oxygen concentration was probably only ~1%. These changes in respiratory gases are correlated with the increasing production of CO_2, and increasing consumption of O_2, by the eggs. The close correlations between CO_2 production and CO_2 levels in the nest strongly suggest that the nest conductance to this gas remains the same throughout incubation.

Most work on gas composition of nests has been carried out in enclosed nests, holes or burrows. Nest CO_2 levels rise to ~2% in the nest holes and burrows of swallows, bee-eaters and woodpeckers although lower values have been recorded in nests of burrowing owls and auklets. The movement of the bird in and out of the tunnel leading to nest chamber has been shown to be very important in ensuring that the gaseous environment in the nest remains acceptable for the eggs, chicks and incubating birds.

Not surprisingly the gaseous conditions in the mound nests of megapodes are very poor (See Box 5.2, p. 79). Values of 12% CO_2 and only 7% O_2 have been recorded in brush turkey nests.

The paucity of data on the gas composition of nests is disappointing. Hopefully, future research will attempt to collect more information on this important aspect of incubation.

Egg position and turning

As was shown earlier (see p. 24) egg shape varies considerably in birds. The shape and number of eggs in a nest tends to affect how the bird actually sits on the nest. A clutch of typical asymmetrical eggs, i.e. with one end more pointed than the other, will tend to form a bowl with the pointed ends at the centre. More rounded eggs would simply form a clump. However, in birds like gulls the eggs are positioned individually so as to fit on the brood patch. In reality, most eggs are simply constrained by the nest and are rarely incubated in any one particular orientation.

Egg position does change with time. At laying the egg is full of yolk and albumen and will adopt a neutral position. i.e. the long axis of the egg is nearly horizontal (Figure 5.17). However, as the egg is incubated and loses weight by water loss from its eggshell pores, the air space grows and so the one end (usually the blunt end) of the egg gets lighter than the other and so the egg naturally adopts a tilt with the air space upwards (Fig. 5.17). This is crucial for the developing embryo because it orientates itself to the top of the egg for the hatching process (see p. 59–62) and so is able to position its beak next to the air space. Without this natural asymmetry in the egg there would be a danger of the embryo developing with its head in the wrong end of the egg, a

Figure 5.17. The effect of development of the air space on the orientation of an egg.

Before development   **At the end of incubation**

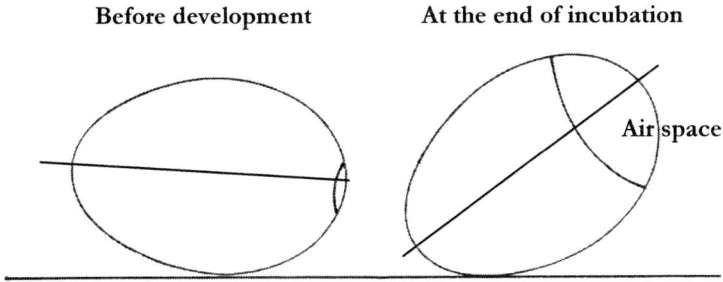

Air space

problem often seen in artificial incubation (see p. 176–178). The adult bird plays no part in this process.

Birds, forced by their role during incubation to sit upon the egg, are very restricted in the behaviours they perform on a daily basis (see pp. 69–71). They normally just sit there, sometimes preening, often keeping an eye out for trouble but mainly just dozing off. Every now and again they do deliberately get up in order to poke around within the nest and by doing so they turn the eggs.

Turning (or "shifting" or "jabbing") the eggs is a rather haphazard procedure. The bird does not pick the egg up with its bill and rotate it before replacing it in the nest cup. Typically turning involves moving the bill around within the eggs often pulling an egg towards the breast of the bird and thereby displacing the eggs around it. Some birds with short bills also use their necks in conjunction with the bill to help move the eggs. It is possible that some species also use their feet to turn the eggs but this is more difficult to confirm (the bird doesn't have to get up to do it). After turning the bird usually settles down again to carry on sitting on the eggs. In some birds with more intermittent incubation, and in birds which share incubation (see pp. 67–68), the eggs are turned when the bird returns to the nest to resume an incubation session.

Egg movement during turning is a random action. Not all eggs move during a turning event and the angle that an individual egg is turned is highly variable. Research has shown that mallard ducks move the eggs by an average angle of turn of ~60° although there was wide range of angles recorded. This seems to be typical of those species (domestic fowl and some birds of prey) where turning angle has been recorded.

So why do the birds bother? Surely, getting off the nest will cool the eggs? For many years the role of egg turning was considered important in redistributing heat within the egg and between the eggs in the nest. However, eggs need to be turned even in the uniform environment of an incubator which rather excludes this possibility. More typically turning has been thought to prevent the young embryo from sticking to the inner surface of the shell membranes. Not surprisingly this was considered to prevent normal development and cause premature death. Research into the role and practise of turning was largely restricted to commercial poultry and was carried out in the laboratory. Hence, modern incubators have adopted a system of egg turning based on this research. However, during the 1980s there was renewed interest in the effects of turning on embryonic development, typically investigated by not turning eggs of domestic species during incubation.

Failure to turn eggs does indeed have high early mortality with embryos stuck to the shell but most embryos survive to develop up to the end of incubation. Unfortunately hatchability of unturned eggs is greatly reduced and most embryos died in the last few days of development. Lack of turning during a critical period (3–7 days in the domestic fowl) can cause all of the problems of unturned eggs even if the eggs are turned at all other times of incubation. It is not known whether all bird species have this critical period.

BOX 5.6 – THE PHYSIOLOGICAL EFFECTS OF FAILURE TO TURN EGGS.

Work on the failure to turn eggs of the domestic fowl and Japanese quail has revealed that turning has quite profound effects on the development of the embryo. It has proved more complicated than the embryo simply sticking to the inner eggshell membrane.

In domestic fowl there is a critical period, between 3–7 days of incubation, during which eggs can be turned and they appear to develop normally irrespective of the turning they receive that other times of the incubation period. Failure to turn during this critical period alone, is as bad as not turning at all.

Development during the first week of development is retarded by failure to turn eggs. The growth of the blood vessel rich *area vasculosa* of the yolk sac membrane is restricted and there is less transfer of salts and water from the albumen into the yolk to form sub-embryonic fluid (see p. 56). These restrictions have knock-on effects on fluid balance later in development. The lack of movement means that the embryo also tends to sink forming a depression in the top of the yolk sac effectively cutting itself off from the albumen.

During the second half of incubation there is less allantoic fluid and the movement of albumen proteins into the amniotic fluid is severely restricted. The chorio-allantoic membrane, which the embryo uses to breathe is also smaller. All these factors combine to slow the rate of embryonic growth to the point that at the end of incubation the embryo is too small to hatch.

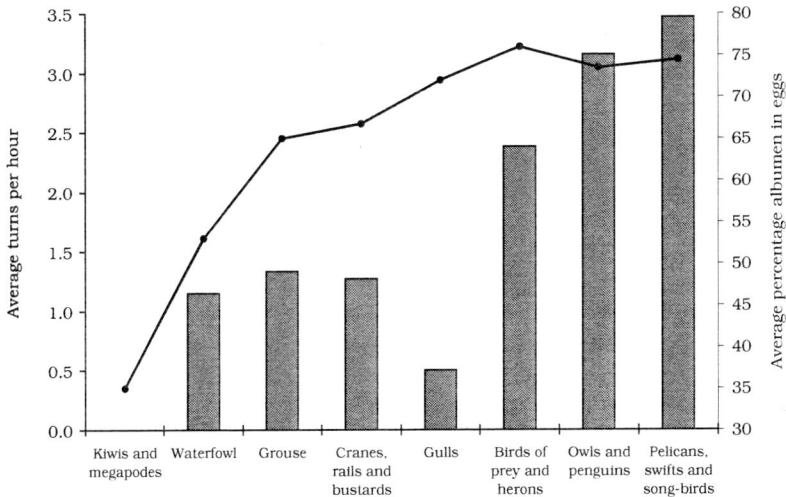

Figure 5.17. Results of research showing that albumen content of eggs (line) is correlated with rates of egg turning (columns).

Lack of turning has lots of physiological effects (see Box 5.6, p. 96) but mainly retards embryonic growth, restricts growth of the extra-embryonic membranes, disrupts the fluid balance of the embryo and prevents normal utilisation of the albumen. Unhatched embryos at the end of incubation are characterised by their small size, small chorio-allantoic membrane and residual albumen lying at the bottom of the egg. The basic problem with insufficient turning appears to be related to the uptake of albumen; experimental removal of albumen from turned eggs at 3 days of age mimics the effects of not turning the eggs.

It is proposed that turning was important to ensure that the embryo could utilise all of the albumen within the normal incubation period. If this is the case then it is also predicted that those species of bird with eggs containing different amounts of albumen will have different rates of turning. Eggs from altricial species have a relatively high albumen content compared with eggs from precocial species (see p. 39) and should have higher rates of turning. This prediction has been confirmed recently by a study looking at reports of the rates of egg turning by birds on nests. These are typically recorded by direct observation of birds although telemetric eggs (See Box 5.7, p. 98) have been used for some species with larger eggs.

The results are summarised in Fig. 5.17. Large eggs of megapodes and kiwis are energy rich with large yolks and relatively little albumen and are not

---

BOX 5.7 – TELEMETRIC EGGS AND OTHER MEANS OF MONITORING THE NEST ENVIRONMENT.

One relatively modern method used to monitor natural incubation is the electronic telemetric egg. These gadgets are designed to be placed and accepted in nests where they record the environmental conditions under the bird. Packed with electronics these eggs contain sensors which typically record the temperature under the bird as well as position sensors that can sense the orientation of the egg and record when its position changes when it is turned. The data recording is usually transmitted by a radio link to a computer for analysis. These systems have the advantage of being self contained allowing the bird to perform normal behaviours in a natural nest. Unfortunately, the electronics mean that the eggs have to be relatively large and are only suitable for species that can accommodate them. To date, telemetric eggs have been mainly used for cranes, penguins and larger waterfowl.

Other electronic eggs have been employed but they have tended to rely on wires for data collection, which force the eggs to be fixed in position. This makes them useless for recording egg turning behaviours.

Other artificial eggs, often perforated with holes have been used to measure nest humidity and gaseous environment but again these eggs tend to be fixed in one position. A particularly useful technique is the use of an egg which has been 'calibrated' – the porosity of the shell has been previously determined in the lab. Placing one of these eggs in a nest allows its weight loss to be recorded under the bird. Because the porosity is known, the nest humidity can be calculated. If weight loss was recorded in the other eggs in the nest at the same time then this allows shell porosity to be calculated for the 'unknown' eggs. Other artificial eggs have been designed to be infused with water at different temperatures thereby allowing investigation of the incubation behaviour of the sitting bird.

---

turned at all during long incubation periods. Eggs of the precocial waterfowl, grouse and Gruiformes (cranes, bustards and rails) have higher albumen contents and are all turned around once an hour. Although the semi-precocial gulls appear to have low rates of turning, albumen-rich eggs from other species are turned much more frequently. Hence, birds or prey, herons, penguins and songbirds all have high rates of egg turning (averaging 3–3½ times an hour). Carla Freed in the US has recorded that the albumen-rich eggs of parrot species have high rates of egg turning commonly averaging 12 turns per hour.

Sadly, despite the apparent importance of egg turning in the development of the embryo there has been relatively little study of this behaviour in nests and so it is difficult to assess the significance of other factors that may affect turning frequency. However, it is safe to say that the rate of egg turning is higher during the day – turning rate drops by about a half during the night in most species investigated. How turning rates change as incubation proceeds is less clear. In many, but not all, species studied turning frequency is higher during the second half of incubation; further work is required to confirm this trend. In cranes, turning rates during dry weather are 2–3 times higher than during periods of rain. In the Houbara bustard, the rate of turning is lowest during the hottest part of the day. More work is needed to assess the impact of the weather conditions on turning in other species nesting in the open.

The vast majority of birds turn during incubation but there are a few bird species which do not turn their eggs. Apart from kiwis and megapodes (Figure 5.17), palm swifts do not turn their eggs because they are glued to the nest. Crested swifts also glue their eggs to the nest and probably don't turn their eggs. Fairy terns nest at the end of branches making life risky for their single egg – such conditions probably prevent turning in this species too. Careful observations are required to confirm these predictions. It is unclear how many other exceptions exist and egg turning behaviour is certainly worthy of more research.

Summary

- Incubation behaviour is controlled largely by hormones

- During incubation birds mainly sit on eggs but attentiveness varies between species

- Attentiveness is affected by factors such as whether both parents incubate, the weather, and egg size

- Egg temperature is maintained by contact with a brood patch

- Egg temperature does not always equate to embryo temperature

- Nest humidity is usually raised above ambient humidity by the bird creating a micro-environment within the nest

- Percentage water loss is largely the same for most species

- The gaseous environment of the nest is changed when the bird stands up or leaves the nest

- Egg turning has many physiological roles in embryonic development

- The rate of egg turning is related to proportion of albumen in the egg with altricial species turning eggs around 3–4 times per hour

6 - Artificial Incubation – Importance and History

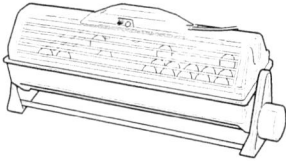

Having gone through the role of incubation in the natural breeding biology of birds over the previous chapters, I now turn my attention to artificial incubation. In this chapter I will address two issues. Firstly, if birds are so good at incubation why do we bother with artificial incubation in the first place? Secondly, I briefly describe the history of artificial incubation from ancient history through to the modern day.

Artificial incubation - why bother?

It has to be admitted that birds are very good at incubating eggs. This may seem obvious but consider the broody bantam fowl. Female bantams will happily sit on any type of egg (size allowing) and are capable of hatching out a wide variety of species. This indicates that the relationship between the bird, the egg and the nest in the bird-nest incubation unit is not exclusive. There are birds that are able to incubate eggs from other species. This ability has been seized upon by brood parasites with great success (Box 5.1, p. 69).

This same principle should apply equally to artificial incubation, *i.e.* use of a machine to provide the environmental conditions that allow normal embryonic development through to hatching. There is no real reason why there are any species of bird whose eggs cannot be incubated successfully in a machine. The difficulty arises in ensuring that the appropriate incubation conditions can be provided. In some species under aviculture artificial incubation has not proved wholly satisfactory and people have used bantam fowl to part incubate eggs. Usually the eggs are incubated for the first half of incubation by the bird and then taken away to complete incubation and hatch in a machine. Whilst this practise certainly works in many cases, I believe that this necessity reflects poorly on our understanding of incubation in individual species and how we try to achieve these conditions in a machine.

A good example of how incubation practise has evolved over time involves the artificial incubation of ostrich eggs. These large eggs (average of 1,500 g) are incubated at 36.0–36.5°C, a full degree lower than that used for poultry or other small eggs. This temperature range was almost certainly been established through trial and error and became accepted because it provided good hatchability. The low temperature does not reflect any idea that ostrich embryos develop at a lower temperature. Indeed the top of the ostrich egg next to the brood patch is 37.2°C (Figure 5.9) and so the embryo is able to develop under an incubation temperature considered normal for other species. However, within the confines of an incubator cabinet maintaining egg temperature at 37.5°C does not provide a good incubation environment for the complete length of incubation. During the second half of incubation ostrich eggs generate and retain a lot of waste metabolic heat (see pp. 83–85). Unless the eggs are cooled efficiently they overheat and the viability of embryos decreases. Maintaining the incubator at 36.0–36.5°C is a compromise between the temperature requirements of the embryo at the start and end of incubation. The lower temperature means that the machine spends more time cooling the eggs. This point was made when I tried to incubate ostrich eggs in a single stage incubator (see Box 6.1, p. 102). The embryos tolerated a set point temperature of 37.0°C for the first few days of incubation before the temperature was decreased to a level more typical for ostrich eggs.

Despite its long history (see pp. 102–108) artificial incubation has often been considered difficult. Most people incubating eggs do not have the luxury of the advantages provided by commercial production of poultry eggs. Here, financial considerations have driven development of high quality incubation equipment and good practise. It is likely that modern strains of poultry are adapted to be incubated under commercial conditions – only those eggs that hatch under such conditions provide birds to lay parent birds for commercial production of meat or egg-laying birds. By contrast, aviculture of more exotic species for conservation purposes, or for the love of keeping and breeding birds, has been reliant on trial and error in part, and on the experiences of the poultry industry, to develop incubators and practise suitable for the eggs involved. Indeed in zoos small incubators often need to be able to cope with a variety of egg sizes and species all at once. In theory artificial incubation should be considered as easy but in practise at times it often seems to be fraught with difficulties. So why not leave it to the birds? Artificial incubation is sometimes a necessity, sometimes the preferred option. The main reasons for this are outlined below (in no real order of importance).

Removal of eggs from the parental nest will often induce the parents to produce more eggs. "Pulling" eggs mimics loss of a clutch through natural predation and means that the number of eggs produced in a laying season can be greatly increased. This has been taken to extremes in commercial poultry production. Having more eggs in a season is advantageous in some species because of their rarity or high value.

Artificial incubation is often a prelude to artificial rearing of the offspring. The sheer numbers of poultry chicks produced each day by a single

---

BOX 6.1 – MULTI- AND SINGLE STAGE INCUBATION

For many years commercial incubation was based on the practise of multi-stage incubation. This involves incubating eggs of different embryonic ages within the same incubator. Usually the eggs from different settings are interspersed within the cabinet in order to produce a relatively homogenous mix of egg ages. In this way there is efficient transfer of heat from the warmer, older eggs to the cooler fresh eggs. Multi-stage setting can be based on any size incubator but is characterised by having a single temperature setting and one humidity setting for the whole machine.

Single stage incubation involves having only one setting of eggs within the cabinet (usually filling it). The eggs are set fresh and are incubated under a programme of temperature settings that change as development proceeds. Hence, the set point temperature is slightly higher than average at the start of incubation but progressively decreases after during the second half of incubation so as to ensure that the eggs are cooled properly. There is usually one humidity setting but this can be modified to suit the humidity requirements of individual settings of eggs. A hatcher is a form of single stage incubator. The difference is that the eggs are not fresh when they are set.

---

hatchery necessitate rearing by humans.

Some parents are poor incubators under captive conditions. Individual birds can be sensitive to human disturbance during incubation and may neglect or even destroy eggs. Their removal to artificial incubation provides a means of hatching these eggs, which would normally be lost.

Sometimes it is not possible for birds to successfully incubate in captivity. For instance, the birds live a colonial aviary and would face disturbance or egg predation from other birds.

Another reason for artificial incubation simply that people are interested in the challenge of incubating bird eggs in a machine. We must never underestimate the satisfaction of taking an egg successfully through to hatching.

A brief history of artificial incubation

The ancient Egyptians are usually credited with the first development of a method for artificial incubation with the earliest mention being in Aristotle's *Historia Animalium* from the 4th century BC. Presumably the Egyptians were interested in increasing bird numbers for food. They constructed large ovens from mud-bricks that they heated with hot air produced from fire of slow-burning materials such as bean straw. The Egyptians also recognised the need to turn eggs during incubation and so they turned them by hand.

The Chinese had also developed artificial incubators at least by the middle of the 3rd century BC. These were "k'angs" or hatching ovens and again the heat from supplied from fires of slow burning material such as dry manure or rice husks. Several k'angs were operated in sequence and eggs were moved from one k'ang to another after the heat was lost from the first. Eggs were turned by hand.

Reports of alternative methods in both locations included using rotting dung to generate heat. Whether these reports have any validity is unclear but they may refer to use of dry manure being burnt to supply hot air in ovens. In China and the Far East heated rice husks were used to surround the eggs. There are also reports of people being employed to incubate eggs by lying on top of eggs laid out in rows on wooden frames in especially constructed "sofas".

The Egyptian and Chinese incubating ovens appeared to have been used to hatch duck eggs in the first instance and then the technology was applied to other poultry. These techniques have persisted to this day and are still used for small-scale production of ducklings and chicks.

The proximity to Egypt meant that their techniques were investigated and applied in Europe. Success for much of the Middle Ages was very limited partly because of the lack of technology but the different climatic conditions probably prevented maintenance of the correct incubation temperatures. However, interest in developing useful techniques for artificial incubation began to develop around the time of the Renaissance. Key events in the history of incubator design and operation are described below and summarised in Table 6.1.

In 1558 Gian Battista della Porta developed an incubator consisting of an insulated wooden box warmed by hot air from a lamp. The eggs were placed in sawdust sharp-end downwards on trays within the box. Around this time Dutchman Cornelius Drebbel invented an incubator with more practicable design. This was notable for its use of water-jacket heated by hot air to provide an indirect, even distribution of heat and the invention of a mercury-alcohol thermostat. Despite these innovations, Drebbel's invention was largely forgotten although it was described by his son-in-law at the Royal Society in London in 1668.

During the 1650s Grand Duke Ferdinand II of Toscana was interested in developing an incubator and used an alcohol thermometer of his own design to monitor the temperature of the Egyptian style oven incubators he constructed. In 1659, two members of the Accademia del Cimento in Vienna investigated artificial incubation and although largely unsuccessful in making any great technological advance, they were the first to use the new thermometer

Table 6.1. Timing of key events in the development of artificial incubation.

Date	Event
4th century BC	Egyptian oven incubators
3rd century BC	Chinese k'ang oven incubators
1550s	Gian Battista della Porta sets eggs sharp point downwards. Cornelius Drebbel designs incubator with water-jacket and thermostat
1650s	Use of a thermometer to monitor incubation temperature under a bird and in an incubator.
1720	Leutmann recognises the role of humidity during incubation.
1749	de Réaumur's book on incubation stimulates great interest in incubation.
1780	Copineau monitors humidity during incubation using a hygrometer.
1816	Bonnemain re-invents the thermostat.
1880	Axford produces first warm air machine with an electric thermostat. Martin patents a hand-operated roller for turning all of the eggs in the machine at one time.
1881	Hearson patents the *Champion* incubator with an ether-alcohol capsule thermostat.
1907	First "Mammoth" incubator.
1911	First force-draught incubator.
1923	First all electric incubator.
1930s	Consolidation of designs of electric incubators. Development of contact thermometer control.
1960s–1970s	Use of new materials in incubator construction
1980s	Development of large-scale single stage incubators.
1990s	Development of computerised controls.
2002	Development of the *Brinsea Products* artificial contact incubator

to determine the temperature under an incubating fowl. They used this information to adjust the temperature of the ovens they constructed. In England Sir Christopher Wren and Robert Hooke were also investigating artificial incubation during the 1660s. Hooke was able to design a self-regulating lamp. Around this time interest in embryology also began in England and France.

Leutmann published a book in 1720 (running to five editions!) in which he described furnaces and ovens that could be used for artificial incubation (Figure 6.1). Furthermore, Leutmann proposed three principles that had to be fulfilled before artificial incubation could be successful: 1) careful study of the temperature under incubating hens; 2) construction of an oven that will hold a fire for 12 hours but which can also be regulated in such a way that tempera-

Figure 6.1. A plan of Leutmann's 1720 oven-type incubator as seen in section from the side. The furnace heated air that passed through a flue into the egg chamber. Temperatures inside the chamber were tested using the "tin egg".

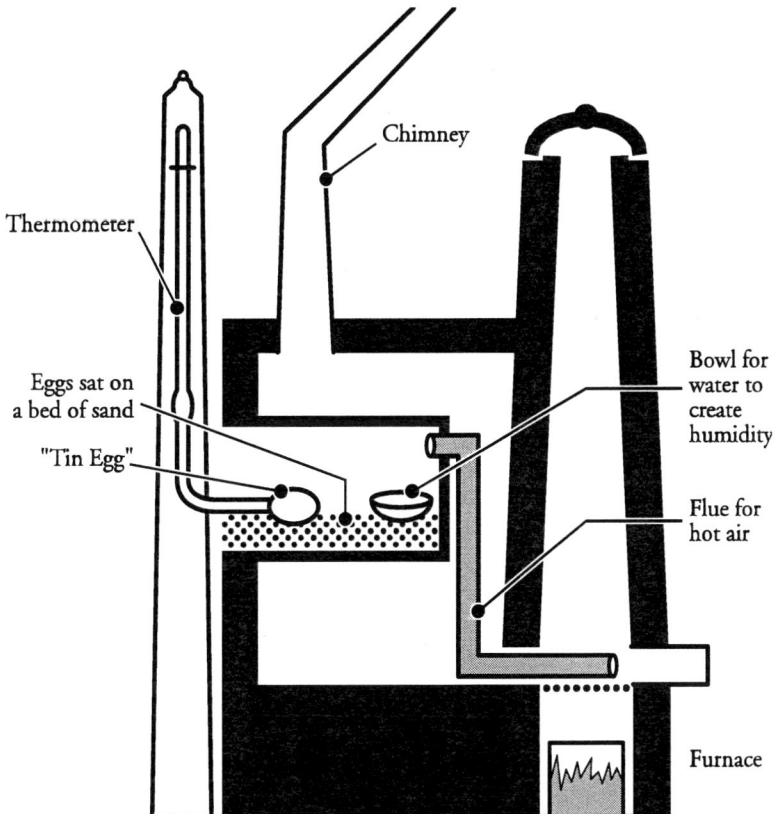

ture of the incubating compartment can be maintained at a level corresponding to the conditions recorded under the bird; 3) adjustment of humidity according to natural conditions. Leutmann constructed a "tin egg" connected by a lead tube to a thermometer to measure temperature under the bird and in the incubator.

In 1749, Beguelin published a book describing an incubator designed to allow embryological study. To facilitate observation, Beguelin removed the eggshell over the blunt end of the egg.

de Réaumur also published a book on incubation in 1749. This proved to be highly influential even though the techniques de Réaumur described were far from being novel. de Réaumur did, however, emphasise the necessity for preventing excess humidity, and monitored humidity by introducing a cold egg into the incubator and examined the degree of condensation on the shell surface. He also devised a means of ventilating the egg chamber and he insisted on the need for egg turning during incubation. One of de Réaumur's greatest achievements was to develop a method of artificial brooding of chicks hatched from his incubators.

de Réaumur's book stimulated tremendous interest in artificial incubation and in particular methods for the mechanical regulation of temperature. Several of these techniques involved using the expansion of air under rising temperature to open vents to redirect hot air away from the incubator's egg chamber. Some of the incubators that appeared soon after de Réaumur's book were of considerable sophistication and size – Huzard built an incubator insulated with wood and designed to hold 6,000 fowl eggs. In 1780 Copineau published a book critical of de Réaumur and provided details of his own incubator of a similar design to Huzard. Importantly, Copineau made regular determinations of humidity of the air in the incubator using a hygrometer invented some five years earlier by Deluc. Even so Copineau only achieved hatches of 20%!

For many years artificial incubation did not prove popular or meet with any great acceptance by the public. During the 1800s the use of broody fowl and turkeys for large-scale incubation emerged in France.

In 1816 Bonnemain advanced the development of incubators greatly by reinventing the thermostat and applying it to artificial incubation. Bonnemain used a hot water circulation system and thermostatic control of the stove door.

Steam was used first for providing the heat for incubation in 1824 and hot water from natural springs was suggested as a source of heat by Felgère in France (and is still used in parts of China today). Bonnes produced an incubator similar to design to that of Bonnemain's idea in 1831, which included humidity water pans that could be filled from outside. Incubators heated by hot water proved popular for many years to come but incubators remained relatively unsophisticated and unreliable.

In the later half of the 1800s the USA proved a hotbed of incubator design. Boyle made an incubator in 1872 that had a sensitive thermostat in which the expansion of water in a sealed tube operated a lever that moved the oil lamp away from the boiler. Unfortunately it proved too complicated and never found favour. In 1880 Axford from the USA filed a British patent for the first warm air machine with an electric thermostat – expansion of a column of mercury completed a circuit that activated a magnetic valve on the warm air ducting. Also in 1880 Martin from France patented a hand-operated roller for turning all of the eggs in the machine in one go. Martin's machine also had an electric alarm system that operated a set of bellows that blew out the oil lamp. Unfortunately there was no automatic way of re-lighting it!

In 1881, Hearson patented his incubator based on top heating from a water reservoir (Figure 6.2). There was a ventilated egg drawer and water-soaked pads to provide humidity. Hearson's thermostat proved to be reliable, accurate, simple to operate and cheap to manufacture. It consisted of two pliable

Figure 6.2. Plan of Hearson's *Champion Incubator* from 1881. Hot air from the oil lamp moved through a pipe through the water reservoir. Once the required temperature was achieved the ether-alcohol capsule thermostat expanded pushing a rod upwards making a pivot arm rise and open the flue above the oil lamp. Hot air then escaped rather than warming the water reservoir.

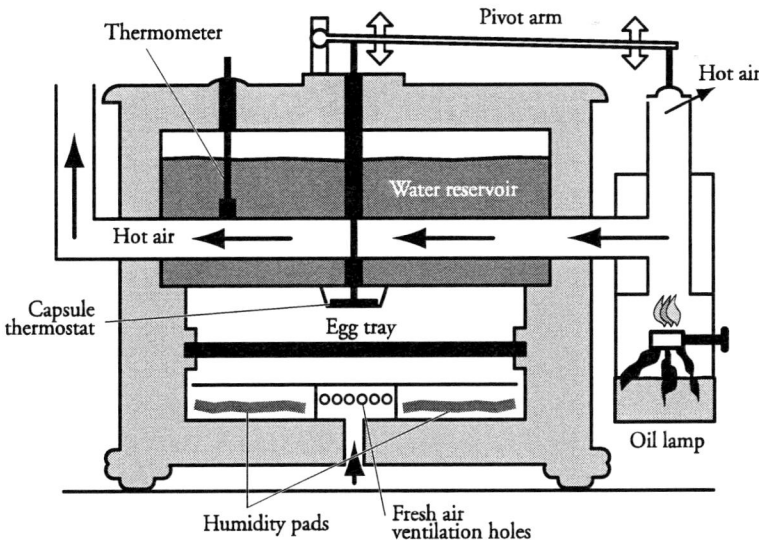

metal plates soldered together along their edges and enclosing a pad soaked in a mixture of ether and alcohol. Expansion of this mixture caused the metal capsule to expand and push a rod that opened the top of the flue over the oil lamp (Figure 6.2). The warm air then escaped rather than passing through pipes in the water reservoir. Hearson also patented a warm air machine in 1883 but machines based on hot water persisted until electric heating took over in the 20th century.

The 20th century saw a massive expansion in incubator design and manufacture. Particular innovations included the "mammoth" incubator constructed by Cyphers in 1907 to incubate 36,000 duck eggs. In 1911 Hastings filed a patent for the first forced-draught incubator (See Box 6.2, p. 109) that employed a fan to move air around the cabinet. In 1923 Petersime developed an all electric incubation system and modern commercial incubators were born.

Over the rest of the 20th Century commercial poultry incubators evolved to become larger in size and more complicated in design. Machines were all force-draught and of large capacities operated as multi-stage machines. The cabinets had insulated wooden walls and there were electrical heating and water cooling systems. Turning was automatic using electrically driven motors. Most machines were of the "fixed-rack" multi-stage design where trays of eggs of one egg age were placed in between other trays of other egg ages. Contact mercury thermometers superseded thermostatic control.

It wasn't until the 1970s that new materials, *e.g.* plastic laminates, metal sheets and fibreglass, began to be used to manufacture the cabinets. Electronic control cards replaced contact thermometer control only to be superseded by computerised control systems, which can be linked to remote computers. Hot water re-emerged to heat large-scale incubators because operating energy costs became of more importance. Development of plastic egg trays and trolley systems meant that egg setting could be done on farm and setting could be done in minutes. Trolley-based multi-stage incubation became the norm and during the 1990s single stage incubation was the preferred option for hatcheries designed to last into the 21st Century.

Commercial poultry incubators have capacities of up to 120,000 fowl eggs but the 20th Century also saw the development of many types of different small-scale incubators. Again wood was the preferred option for cabinet construction for many years and often machines were still-air systems. New plastics and other materials have seen incubators becoming smaller. Electrical heating dominates the small incubator market and many are force-draught machines. Recent innovations include use of flexible plastic sheets containing ink that generates heat when an electrical current is passed through it, and a new innovative contact incubator (see pp. 190–192).

BOX 6.2 – STILL-AIR AND FORCE-DRAUGHT INCUBATION

For most of the years of development of artificial incubators the application of heat to the eggs was via radiation or by convection. The air in the cabinet holding the eggs was not artificially mixed. These "still-air" machines developed to include heating systems in the top of the machine that generated a temperature gradient within the cabinet with the top of the machine being the hottest part of the cabinet. The hot air leaving the cabinet at the top of the box draws in cold air from the bottom. This design means that there can only be one layer of eggs within the cabinet. To increase the capacity of a machine manufacturers often use several separate compartments each with their own heating and control system. Still-air machines are restricted to table-top machines and to multi-compartment hatchers often still used in the game hatching industry.

By contrast, "force-draught" incubators employ fans to move the air around the cabinet. The intention is to destroy any temperature gradients and to ensure that each egg is at the same temperature irrespective of its position within the cabinet. It has been surprisingly difficult to achieve complete temperature uniformity in incubators and many designs still have areas where the air flow is poor and the temperature is either too low or too high. Force-draught incubators can be of any size and in larger machines stacks of egg trays are in trolleys around 2-m tall.

Research into incubation

As was suggested earlier artificial incubators have been used for the study of embryonic development in birds from the 17th Century onwards. Most studies concentrated on the pattern of development in eggs from a variety of bird species. Following the development of commercially viable incubators there was considerable interest in improving results. During the 1930s–1950s there was a lot of applied research into poultry incubation in order to improve hatchability. Once acceptable results had been achieved, interest in research slowly waned and at the present time it is difficult to fund any research into commercial incubation.

During the 1970s–1980s there was a considerable amount of interest in the biology of bird eggs and natural incubation. Much of this work was carried out by Hermann Rahn, Amos Ar and Charles Paganelli and they deserve a lot of the credit for increasing our understanding of the interaction between the egg and its incubation environment in a nest and in a machine. A lot of their discoveries and ideas are described in this book.

Summary

- Artificial incubation is typically used to increase the number of eggs produced by birds

- Incubation in machines it is sometimes necessary to replace birds that are poor at maintaining incubation in captivity

- Artificial incubation has a long history lasting over 2,400 years

7 - How Artificial Incubators are Supposed to Work

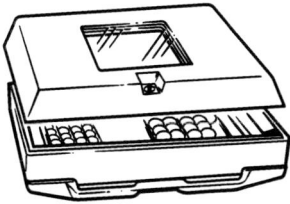

I like to describe incubators as warm, humid boxes with holes in and in which eggs are turned. Oh they were so simple! Incubators come in a wide range of sizes from the smallest table-top machine, perhaps holding only 10 eggs, through to commercial poultry incubators holding up to 120,000 fowl eggs. Each type of machine should be based on the same principles of design and operation with the only differences representing the variation in how these principles are converted into working machines. In this chapter I briefly describe the three basic sizes of incubator: table-top, cabinet and commercial and how these differ. Since most readers are likely to be interested primarily in small table top incubators, these will be a priority with attention to the difference between forced-draught and still-air machines. I then describe the basic working parts of artificial incubators and explain what role these components have during normal incubation. These parts are grouped thus: the cabinet, the system for aeration and ventilation, temperature control, humidity control, and turning. Moreover, I will also show how things can go wrong in the operation of incubators (the consequences of which are described in the following chapter). Finally in this chapter, I consider the factors that need to be considered when deciding on the incubators and hatchers to purchase to suit your individual circumstances.

Essential requirements

In order for eggs to incubate certain key parameters must be met. These are common to all types and sizes of incubators though the method of achieving the same object may be quite different. The data given here is largely obtained from extensive studies into commercial poultry incubation, because this is the only area of artificial incubation that has been studied in sufficient detail. However, much of it is likely to hold good for other species suitably adjusted for egg size. Since the size of an incubator limits the number of eggs it can handle, it is likely that ventilation rates would still be suitable.

Temperature is undoubtedly the most critical parameter because the embryo needs to be maintained within certain limits. Experience with artificial incubation of poultry has shown that machines usually operate best within the range 37.3–37.8°C in a force draught incubator, or slightly higher, 39.0–

39.5°C, measured at the top of egg, in a still-air incubator.

Relative humidity needs to be within certain limits for the embryo to develop properly and to reach hatching time with the right amount of space inside the eggshell to enable it to manoeuvre into the correct position prior to the complex hatching procedure (see Box 5.4, p. 90). The eggs of the domestic fowl and many domestic birds should loose 12–13% of their initial weight, whereas eggs of parrots and birds of prey often need to loose a lot more, 16–18%. It is usually necessary to add moisture to the incubator air to prevent eggs drying excessively.

As the embryo develops it consumes oxygen and liberates carbon dioxide as a waste product. Fresh air must be introduced into the incubator to provide sufficient oxygen and to limit the rise of carbon dioxide to no more than 1.0% (This is 33 times higher than the level in fresh outdoor air). The rate of oxygen consumption and CO_2 production rises sharply in the second half of incubation to a maximum just before hatching. Thus the critical minimum ventilation rate arises near hatching and needs to be at least 1.7 litres per hour per fowl egg weighing 60g. For other eggs fresh air ventilation can be assumed in proportion to weight.

Still-air incubators rely on convection to produce a laminar flow (non-turbulent) of air from the bottom to the top of the cabinet (See Box 6.2, p. 109). A fan (or fans) is often used in other machines and is essential in larger incubators with multiple layers of eggs. The air movement distributes the heat from heaters (and from warmer eggs) to cooler parts of the incubator, thus evening out temperature differences within the cabinet. Second, the air velocity over eggs at later stages helps to slightly cool these eggs; bringing them closer to incubator air temperature. Due to their increasing metabolism eggs liberate significant heat at later stages of development and this has been measured at about 0.14 watts for a 60 g fowl egg. Finally, fresh air vents are usually sited so that the fresh air ventilation rate is influenced by the circulating fan(s). This helps to ensure reasonably consistent rates of air introduction.

Almost all birds turn or move their eggs during incubation (see pp. 95–99). With poultry, turning once per hour works well in incubators and the angle through which eggs are turned is usually 90°. New research suggests that turning frequency should be much higher for eggs of altricial species although the average turning angle recorded in the few nests studied are generally only 60°. Usually only the more modern machines allow for changes in the rate of egg turning.

Types of incubator

"Incubator" is a broad term for cabinets in which eggs are incubated and machines in which eggs are set and incubated in until they are transferred a day or to before hatching, are often called "setters". These machines invariably have circulating fans and automatic turning systems. Incubators into which eggs are transferred just prior to hatching are called "hatchers". They do not have provision for automatic egg turning and are often still-air machines.

Table-top incubators are small in size and egg capacity (usually only 10–100 eggs in a single layer). Access to the eggs is generally from the top by a removable or hinged lid. They are often constructed of plastic or wooden cabinets and have windows or clear plastic tops so as to allow operators to see the eggs inside. The machines utilise electrical heating and do not usually have any cooling system. They often have fans to move air around in the cabinet but many are still designed to be still-air machines (Box 6.2, p. 109). Temperature control is typically electronic but models are still made which use capsule type thermostats (see pp. 107–108). There is often no active control of humidity, which is commonly provided by water pans in the base. The more sophisticated machines do have humidity control systems that deliver water into the cabinet as required. Eggs sit on a perforated or wire base or sometimes in plastic trays. There may be adjustable vents to control the amount of air entering or leaving the cabinet, or the ventilation openings may be of fixed size. Turning can be manual or often automatic using a small geared motor to move the base, a rotating wire frame or by moving the whole machine.

Cabinet machines are usually of a larger scale holding perhaps a few hundred eggs set in multiple layers. Access is generally by a door at the front and the units are free standing. Constructed of metal and or plastic cabinets, resemblance of many of these machines to "chiller" cabinets for drinks is not coincidental. They are all force-draught machines, heating is electrical and many of these machines will also have water-cooling systems. Control of temperature and humidity is usually by electronic systems. Humidity supply may be passive (water pans) or active (sprays). The air inlets and exhausts usually have some means of controlling their size and are manually operated. Egg trays are turned *en masse* by an electrical system. Whereas table-top incubators are often designed as incubators, "cabinet" style machines are sometimes built into cabinets designed for another purpose.

Commercial incubators are, on the other hand, designed as incubators. Large capacities (~20,000–120,000 fowl eggs) mean that the machines are often the size of large rooms and are usually constructed *in situ*. Heating can be from electrical elements or hot water pipes and most machines utilise water-cooling. Those types that rely on cold air alone have specific operating parameters to ensure proper control. Computers control temperature and humidity in the more modern designs although older machines are still around that use electronic control cards or mercury contact thermometers. The machines are all force-draught employing numerous large powerful fans. In many incubators damper flaps altering the size of air intake and exhaust holes are under computerised control. Turning is achieved by electrical or pneumatic rams for whole banks of egg trays held in racks or on trolleys. Note that the larger the incubator the more necessary it becomes to provide cooling. Commercial incubators hold thousands of eggs all generating heat and their low surface area to volume ratio means that they cannot loose enough waste heat without additional cooling.

Construction and operation of incubators

The perfect incubator has not yet been built. In this section I will describe the principles of operation of various components of the machines. In this way it is hoped that the operator will gain a better understanding of how machines are supposed to operate.

Cabinet

Eggs are kept within a cabinet typically constructed of wood, a composite of metal sheets insulated with expanded polystyrene, or plastic material that insulates the contents from the outside air. Any windows in the walls or top of the machine need to be double-glazed to maintain the insulation. The cabinet's role is to create and maintain a micro-environment for incubation. It is certainly easier to control the environment of the smaller volume of air within the cabinet than that of a room. The cabinet also has holes for fresh air ventilation and provides support for the heating elements, egg tray racks, turning mechanism and where appropriate, cooling pipes.

The cabinet cannot work in isolation and the location of the incubator can be crucial in maintaining and controlling the correct incubation environment. The temperature of the room in which an incubator is kept is important. If either too cold or too hot then the machine will struggle to maintain the correct incubation temperature. This is particularly important for small table-top machines that should be kept in a room around 20–24°C. Any cooler and the heat loss from the cabinet walls may be excessive preventing the machine from maintaining the correct internal temperature. In small machines heat loss through the cabinet walls is crucial for the heat balance within the box. In warm rooms there can be insufficient cooling through the walls and the air entering the cabinet is warm and has a negligible cooling effect. Under these conditions eggs can overheat. Generally, well insulated machines are energy efficient and are likely to have less temperature variation within the cabinet but they must not be used in high ambient temperatures approaching incubation temperature. For all sizes of machine the room should act as a coarse control for the incubator and the better the control of the room environment the better.

For instance, eggs from desert-nesting birds have shells with low porosity to counteract the low humidity environment in the natural nesting site. Taking these eggs into an incubator means that a low humidity is required in the cabinet to get the correct weight loss during incubation. This will be difficult to achieve if the room is very humid. To achieve the necessary low humidity ostrich farmers around the world have to rely on de-humidifiers to remove humidity from the room before it enters the cabinet.

Ventilation and aeration

Still-air incubators (Box 6.2, p. 109) rely on convection to move the air through the cabinet. Warm air rises and escapes from the cabinet through any holes high up in the walls or roof of the machine. This movement of air pulls in cool fresh air from any holes near the bottom of the cabinet (Figure 7.1). This flow of air by convection ventilates the cabinet. However, the rate of fresh air ventilation is in direct proportion to the difference in temperature between the air inside and outside the box. Thus, as the room temperature rises closer to incubation temperature, the air flow reduces, so in hot ambient conditions there is a risk of inadequate fresh air supply as well as the possibility of overheating.

In many large scale still-air hatchers used for game birds the fresh air enters and leaves the cabinet at the back of the machine. The positioning of the

Figure 7.1. Diagrammatic representation of the air movement through a still-air incubator as seen from the side. A temperature gradient (indicated by the grey triangle with the hotter temperatures at the top of the machine) is caused by the heating elements at the top of the cabinet. The temperature probe is located at the top of the egg. Warm air rising moves out of holes at the top of the cabinet and draws in fresh air in through holes in the bottom of the machine.

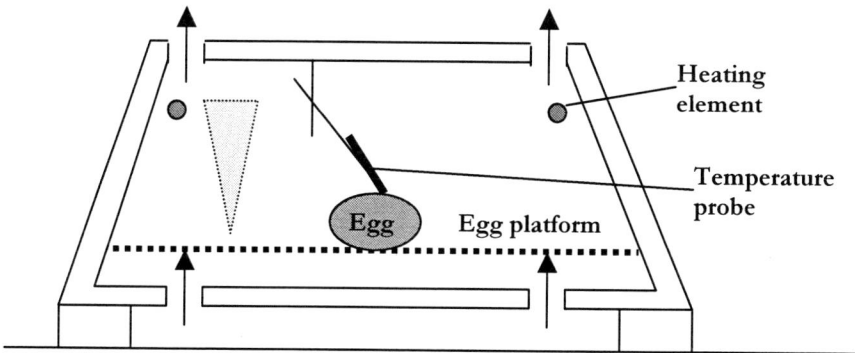

hatcher basket tray is crucial. It has to be flush against the back wall of the cabinet so that air entering the cabinet is forced to move forwards underneath the hatcher basket (which has a solid base) and over the humidity pan. The cool humid air then moves upwards by the door at the front of the machine and then towards the back of the machine passing over the layer of eggs before leaving through the exhaust holes at the back of the machine. If the hatcher basket is not flush to the back wall of the cabinet then fresh air bypasses behind the hatcher tray rather than moving through the cabinet.

Maintaining a uniform temperature profile within the incubator is critical for optimising the rate of development for all embryos and restricting the spread of hatch times. In force-draught incubators a fan, or fans, within the cabinet stirs up the air moving it around between the heaters, coolers and eggs. It also mixes the fresh air with the cabinet air as well as the water vapour generated by the humidity systems. Hence, in force-draught incubators the fans play a critical role in providing a uniform incubation environment around the eggs.

Fan speed is very important. Medium-sized "barrel" setters have a fan rotating around the bank of egg trays and experiments have shown that the faster the fan goes the better the hatchability. This is almost certainly due to reduced temperature differentials among different eggs. In the large commercial incubators the speed of the air is important in determining the temperature difference between the eggs and the air passing over them. The faster the air then the lower the temperature difference. Unfortunately, the faster the air moves the more turbulent it becomes. Placing obstructions in the path of the air movement doesn't help matters. The size, speed and location of fans, and the configuration of air passages between eggs interact in very complex ways, which are difficult to predict. Much experimentation goes into the design of good forced-draught incubators.

The ventilation fan is also usually responsible for the aeration of the incubator. The movement of air past a hole in the cabinet usually pulls in fresh air from outside of the cabinet. The relationship between the moving fan blades and the hole is important: the shorter the distance then the more air will be pulled into the machine. Many incubators have damper flaps that can be used to adjust the size of the hole. Here the amount of air entering the cabinet is controlled by the size of the hole but the maximum amount of air being pulled in is ultimately a function of the positioning of the fan and the hole.

It is essential to have at least two vent holes in an incubator for fresh air to be drawn though. The pressure difference generated by the circulating fan can only draw in fresh air if there is also a way out. I know of one brand of small incubator that has just a single hole and can cause asphyxiation of embryos unless modified. A further surprising problem can arise when people modify

Table 7.1. Summary of the possible sources of heating and cooling within incubators. Maintenance of any particular depends on balancing heat input with heat loss.

Heating	Cooling
Heaters (Electrical, or sometimes in large machines, hot water)	Heat loss through the cabinet walls (particularly important for small incubators)
Metabolic heat production by eggs towards the end of incubation	Background cooling from fresh air continuously entering the cabinet
Friction between fan blades and air, and heat from motors in the cabinet	Cold water coolers or fans that blow cold air into the cabinet
	Absorption of heat energy by cooler fresh eggs

still-air incubators. One brand of machine works perfectly well when used as it was designed, *i.e.* without a fan. Adding a circulating fan positioned in the centre of the top of the cabinet causes air to blow across the exhaust vents, which tends to pull in fresh air. However, in the design of the product these holes were arranged for convective flow of air out. The consequence is that almost no air enters or leaves the cabinet with dire effect unless other steps are taken.

In all types of machine there has to be movement of air through the cabinet for maintenance of the appropriate incubation environment. The amount of air entering an incubator cabinet is important for the provision of oxygen-rich air. Exchange of the air within the cabinet is crucial because it prevents build up of humidity and carbon dioxide (CO_2) and a reduction in oxygen. High humidity can restrict weight loss from the egg and adversely affect hatchability (see pp. 145–147). Embryos of all ages are tolerant of high levels of CO_2 (over 1% of the air) at most times during their development (see pp. 167–168) but lack of oxygen is much more serious. Although the amount of oxygen required by the eggs may be provided by a relatively small amount of air it is nonetheless crucial for normal development and growth (see pp. 169–171).

Temperature maintenance

For all types of machines temperature control is a matter of balancing heat loss and heat gain of the cabinet in order to maintain the correct temperature for the eggs. There are relatively few sources of heat, and means of cooling eggs, during incubation and these are outlined below and summarised in Table 7.1.

Electrical heating elements provide the primary source of heat energy in small incubators. These are usually resistance wire heating elements although they can be special flexible plastic sheets encapsulating conductive ink that produces heat under an electrical current. The advantage of the latter system is that it creates a uniform heating surface

Developing eggs provide the second important source of heat in a cabinet. As the embryos develop and grow they produce increasing amounts of heat as a waste product of normal cell metabolism. During the last third of incubation, and particularly in hatchers, this heat output can be sufficient to raise the temperature of the egg a measurable amount above that of the cabinet air.

Fan blades within the cabinet also generate some heat through friction with the air. A fan usually makes little contribution to the heat load of an incubator, but any electrical motor located within the cabinet will also get warm and heat the air. The importance of this depends on the type of motor and the size and construction of the incubator. Many small incubators in the USA are made of polystyrene foam with a high insulation factor and were originally designed as still-air incubators. A "turbofan" modification is available from two manufacturers to convert them to forced-draught machines. However, the fan and motor are both fitted inside the insulated cabinet and the kind of motor used generates considerable heat, perhaps one third that of the heater. The consequence is that such machines are prone to overheat unless the room temperature is kept quite low.

Cooling in an incubator can be achieved in a variety of ways. In small table-top incubators heat loss through the cabinet walls is crucially important in maintaining the correct temperature. Small machines have a large surface area to volume ratio and even though the walls are insulated there is a constant loss of heat. Add to this the point that small machines contain few eggs to produce metabolic heat and the need for supplementary cooling systems is lost. As the size of the cabinet increases the number of eggs increases yet the surface area to volume ratio goes down. Large machines therefore find it hard to lose sufficient heat through the walls and so other cooling systems are required. The walls of small incubators are analogous to the cooling coils of a commercial machine.

The intake of fresh air into the cabinet causes continuous cooling (provided of course that the temperature of the air entering the cabinet is lower than in the cabinet). This cooling in many table-top machines is usually less than 10% of the total cooling experienced by the system. In some large commercial incubators fresh air entering the cabinet is more important as a cooling system and dampers are used to control the amount of air entering the cabinet. Some commercial incubators and particularly hatchers have auxiliary fans that can blow in vast amounts of fresh air into the cabinet as required. This cool air can be a mixed blessing – if the air unduly influences the temperature sensor then maintenance of the correct incubator temperature can be compromised (see Box 7.1, p. 119).

In many commercial machines water pipes for cooling are important components. Cooling using water is very effective because it absorbs heat from the air over a relatively long period of time, thereby assisting in the effective removal of metabolic heat.

The eggs themselves are important source of cooling in multi-stage incubation (Box 6.1, p. 102). Placing fresh eggs into a machine containing eggs already under incubation provides a "heat sink" for metabolic heat generated by the older eggs. This transfer of energy is only efficient if the eggs of different developmental are mixed up within the machine and there is an efficient ventilation system.

Not all incubators have all of these systems. Hence, a small table-top still-air machine will be heated by electrical elements and the eggs, and cooled mainly by loss of heat through the walls, fresh air entering the machine and by placing cold eggs into the cabinet. By contrast, a commercial hatcher will be heated by electrical elements, the eggs and friction, and will be cooled by water pipes, cool air entering the cabinet as well as having an additional fresh air blower.

Control of incubator temperature is achieved by a control system that measures the temperature of the air circulating around inside the cabinet and turns heaters (and coolers in large machines) on and off accordingly. Artificial incubators have to measure the temperature of the air and assume that this is a close representation of the egg temperature. Unfortunately this is not always the case.

Monitoring temperature

Nowadays temperature sensors (see also Box 7.1, p. 119) are usually thermistors although some machines still utilise contact mercury thermometers. Large numbers of low cost machines still use ether capsule thermostats. Thermistors are resistors whose value changes rapidly with change in temperature, and are connected directly to the control system of the machine. This is usually electronic but computerised systems are common in modern commercial incubators and in a few sophisticated smaller machines.

Electronic controls in modern small incubators are usually "burst fire, proportional" systems. These systems switch the heater supply on and off constantly for periods of a few seconds. The ratio between the "on" periods and the "off" periods is constantly varied in proportion to the measured departure from set temperature. With careful attention to design of the complete system, control accuracy comparable with commercial machines can be achieved in small machines. Reliability can also be very high if they are adequately protected from supply line spikes – the downfall of many early systems. Unfortunately, not all manufacturers have taken this care. Another problem is the use of a common sensor to serve as control, alarm, and thermometer. Many digital systems are like this and of course, if the thermistor should fail or go out of calibration, all systems go down with it.

By contrast, mercury contact thermometers are relatively unsophisticated but hence theoretically more reliable. The position of the mercury in the glass thermometer is used to complete the electrical circuits for heater operation. Each set of contacts is set at different points up the column, hence at different temperatures, and can be used to operate heaters, coolers and alarms. The disadvantage of contact thermometers lies in that fact that if the set-point has to be changed then the whole thermometer has to be replaced.

In an attempt to better understand how temperature is maintained in a machine let us consider the events happening after setting eggs in firstly a still-air, table-top incubator and secondly a force-draught table-top incubator.

Lifting the lid on a still-air machine releases the warm air within the cabinet and the presence of cool fresh eggs lowers the temperature further. Therefore, the heater comes on once the lid is replaced and stays on until the temperature approaches the set-point. In these machines the sensor that measures temperature is usually placed at a level equivalent to the top of the eggs (Figure 7.1). Still-air incubators generate temperature gradients because the heating elements are usually in the top of the machine. Air at the top of the cabinet will be well above the set point but the temperature is lower at the level of the eggs. In this way the still-air mimics the temperature gradients recorded in bird eggs being incubated by a bird (see pp. 82–85). Once at the set point the time that the heater is turned on depends on the heat requirements of the cabinet and the eggs. Cooling is caused by continuous loss of heat through the cabinet walls and by hot air moving out of holes in the top of the cabinet and being replaced by cooler air entering at the bottom of the machine (Figure 7.1). After a time a steady condition is achieved between the heat input and output from the cabinet, the heater being switched on and off for intervals to maintain this balance. Metabolic heat generated by the embryos is lost to the surrounding air and by the end of incubation the eggs may be contributing a significant amount of energy to the air temperature, rather than it coming all from the electrical heaters.

With still-air incubators the positioning of the temperature sensor is important in getting the appropriate egg temperature. Ideally the sensor is at the level of the tops of the eggs. If it is not (as with many capsule controlled models) changes in temperature profile due to external temperature variation will change the temperature at egg level.

Mixing up eggs of different ages within the cabinet can assist in evening out temperature differences. If a still-air machine is used as a multi-stage machine and is full then it is good practise to place cool eggs next to older eggs to produce a more uniform temperature distribution within the cabinet. Of course there can only be one layer of eggs in these machines. Larger game hatchers that use the still-air principle (see p. 115) are actually cabinets that contain a large number of individual compartments each with its own control system.

A force-draught, table-top incubator the difference is that the fan moves the air around the cabinet and the thermometer does not have to be placed at the top of the eggs. This is because there are no temperature gradients in either the machine or the egg. Therefore, the thermometer is placed in the moving stream of air and it is assumed that the air moves around the cabinet in a uniform way. It is also assumed that heat exchange between the eggs and the air is efficient so the difference between the two values is minimal. Once the heater has warmed the air to the set point then the heaters will be on for only short periods and the machine relies on cooling through the cabinet walls and from fresh air to maintain a suitable egg temperature.

Some of the larger table-top machines do have flaps controlling the amount of fresh entering the cabinet so that cooling can be increased or decreased according to the number of eggs in the machine. For instance, if the machine is set with only a third of its total capacity then the amount of fresh air entering the cabinet can be reduced to around a third of the maximum. The position of the temperature sensor is important – it should not be placed in a position where the fresh air entering the machine can influence it.

However, as the capacity of the incubators increases to a commercial scale the heat load generated by the metabolism of the embryos is more critical in the temperature balance of the cabinet. Hence, the ability to accurately record air temperature as influenced by the eggs is critical to maintenance of the proper incubation environment. In some setters incoming fresh air dictates the position of the temperature probe and the temperature set point has to be set at a slightly lower level to take this into account.

Most commercial machines operate in the same way as table-top force-draught machines. They have a heaters-off set point, usually between 37.4–37.6°C, below which heating is applied. They also have a coolers-on set-point some 0.3–0.5°C higher. Typically the heaters and coolers are located in the centre of the machine and immediately next to the circulating fan. Ideally the temperature sensor should be placed well away from the inlet for incoming fresh air. In this way the cold air does not affect the operation of the sensor, which should be registering the air temperature after it has passed over the egg trays. In these machines, the spray nozzle has also to be placed in a position where the water cannot easily affect the temperature sensor.

Humidity supply

A humid environment is a key aspect of successful incubation (see pp. 145–147). Humidity is basically the distribution of water vapour (gaseous water) within the air. The amount of water vapour in the air is usually described as a percentage relative to saturated air (100% relative humidity [100%RH] is the same as saturated air). Temperature influences the amount of water vapour that the air can hold and so for any particular relative humidity, the absolute amount of water vapour is different at different temperatures. For example, air at 20°C room temperature and 50% RH holds 7.38 grams of water vapour per kg of air. The same amount of water vapour in air at 37.5°C incubation temperature would have relative humidity of only 18%. To achieve 50% RH at incubation temperature, the air must carry 20.40 grams of water vapour per kg of air, almost three times as much. Thus although air entering the incubator always carries some moisture, the net effect of the fresh air is that it always reduces the incubator humidity. This loss usually needs to be balanced by water vapour introduced directly into the cabinet.

The eggs themselves also contribute water vapour to the incubator air. Water vapour leaves the eggs during incubation and enters the cabinet air to contribute to the humidity of the air. If the cabinet is fully closed then the humidity of the incubator can rise quickly as water vapour leaves the eggs. This can be considerable – 50 fowl eggs will lose around 20 g of water vapour each day!

It is important to realise that it is not practical to remove humidity from incubators except by increasing the fresh air exchange rate. This works better if the room air is dehumidified already.

Increases in humidity are often achieved by evaporation of water from pans placed within the cabinet. This system depends on the operator topping up the pans periodically. The surface area of water determines the rate of evaporation and hence is a crude control of humidity in the incubator.

Other more sophisticated systems to supply humidity are based on a monitoring of humidity either by electronically sensing wet bulb temperature or by measuring humidity directly (Box 7.2, p. 123). The information from the sensor is used to pump water (or water vapour) into the incubator to maintain the level selected. Such systems are available for all sizes of machine but usually only work well with incubators with fan circulation.

BOX 7.2 – MONITORING HUMIDITY

There are two main ways of monitoring humidity in order to control a humidity supply system such as a spray. The first is to use an electronic sensor that registers humidity directly in the air and usually displays it as relative humidity (see p. 122).

The second system involves monitoring of "wet-bulb" temperature. The "bulb" of a thermometer is the part containing the component that measures temperature – the mercury or an electronic sensor. A "dry-bulb" thermometer simply reads the temperature of the air. Placing a wet cotton wick (connected to a water reservoir) over the bulb creates a "wet-bulb". As water evaporates from the cotton wick there is a drop in temperature compared to the "dry-bulb". The drop in temperature is precisely related to the humidity of the air. A small drop in "wet-bulb" temperature means that only a little water has been able to evaporate from the wick and so the air is very humid. When the air is dry more water can be lost from the wick and the drop in temperature is larger. The relative humidity is determined from a psychrometric chart or tables both of which describe the relationship between wet and dry bulb temperatures.

A third method of measuring humidity is the hair-hygrometer but this is rather insensitive, often inaccurate and cannot be connected to any control system. It does nevertheless give a quick indication of the humidity within an incubator cabinet.

Whatever humidity system is used in an incubator it has to be ensured that water from the humidifier does not come into contact with either the "dry bulb" thermometer (Box 7.2, p. 123) controlling air temperature or the eggs themselves. Wet surfaces cool as water evaporates from them and a wet "dry bulb" thermometer will falsely read the air temperature leading to heaters being turned on and potentially over-heating of the eggs. Wet eggs also cool down and this can slow development and delay hatching.

Egg trays and turning

Egg turning (see pp. 95–99) involves rotating the eggs on a regular basis. In most commercially available incubators there is an automatic system for egg turning. Only in a few of the cheaper machines need the operator to turn the eggs by hand. In this case, the eggs should be carefully rotated through 180° at least three times a day throughout incubation. For the eggs of precocial species turning by hand is not necessary during the night. It is unclear whether this is true for eggs from semi-altricial and altricial species. In nests it has been shown that turning does continue during the night in a few species studied but it is unclear whether this is critical for development. Hatchers do not have turning mechanisms and the eggs do not need to be turned during the last few days of incubation.

Smaller incubators approach egg turning in a variety of ways. Some machines restrain eggs blunt-end upwards in a square mesh frame with the sharp-ends resting on a metal base plate. The mesh is then moved in a circular motion making the eggs sway within the frames. This system relies on the eggs being free to move within the mesh (big eggs tend to get stuck) and have contact with the base plate to provide a pivot point. Other small incubators use rods to restrain eggs lying on their sides on a metal plate or plastic belt. The rods are placed across the egg tray perpendicular to the direction of travel of a moving base plate and prevent the eggs from moving sideways when the base plates moves. The eggs roll as the base plate moves beneath them. Yet another solution to egg turning is for eggs to be set in trays (blunt-end upwards) or lie on their sides in restraining troughs and for the whole incubator to turn.

In larger machines plastic trays are used to hold eggs but these tend to restrict the size of eggs that can be accommodated within the machine but they do maximise the number of eggs that can be set at any given area. Eggs are set blunt-end upward (Box 7.3, p. 125) and the whole tray tilting from side to side achieves turning.

In all machines the typical angle of turn is 45° either side of the vertical (*i.e.* a total of 90°) though whether incubators always achieve this angle is questionable. It doesn't really matter for poultry though because so long as the eggs are turned through at least 35° either side of the vertical (*i.e.* a total of

70°) and once an hour, then they will get sufficient movement. It is important to realise that individual turning events in nests occur in a random manner but in incubators they are regularly alternate in direction. Although this turning pattern is easier to engineer, there is no known deleterious effect of turning in alternate directions and experiments have shown that turning eggs in only one direction is highly damaging to the embryo causing twisting of the extra-embryonic membranes.

Some machines have discrete turning events – the eggs are moved every hour in a short period of time. Although turning frequency is typically once an hour, some more modern machines provide scope for turning rates to be increased or decreased according to the species of bird involved. Many small incubators move the eggs all of the time albeit very slowly. Which method is preferable is not known but for some eggs turning all of the time suggests that in an hour the egg actually moves twice tilting one way and then returning to

---

BOX 7.3 – SETTING EGGS IN THE CORRECT ORIENTATION

There are two ways of setting eggs: 1) they are fixed in one position; and 2) they are free to move and can adopt any position. The position of how the eggs are set is important in determining the correct orientation of the embryo during development.

During hatching the embryo has to orientate its beak towards the air space in such a way that it can break through the inner shell membrane during internal pipping (see pp. 59–61). To achieve this the embryo always orientates towards the top of the egg and in most cases succeeds in achieving the correct position for hatching. Hence, if an egg is set upside down with the blunt end downwards then the embryo will develop a hatching position in the small end of the egg with its beak well away from the air space, which has formed at the other end. Hatching from this position is not impossible but is very much harder for the embryo as it requires external pipping to happen before the bird can fill its lungs (see pp. 59–60). Eggs set blunt end downwards have a much lower hatchability than eggs set normally.

For this reason, if eggs are set in a fixed position within a tray then they must be placed with the air cell upwards. If the egg lies on its side on a base plate then it should be free to move around as it is turned. In this way as the air space forms at one end of the egg then that end will become lighter and the orientation of the egg will naturally shift so that the air space is at the top of the egg (see pp. 94–95).

its original position. Given the recent findings on the rate of egg turning in nests (see pp. 97–98) this turning pattern may benefit eggs from those species producing altricial young.

It is clear from study of turning events in nests that the frequency and angle of turning are carried out by birds in very much a random manner. Time intervals between turning events, and which eggs are moved and by how far, are not fixed. Only in species where the local environment constrain incubation behaviour is there any pattern of egg turning. For instance, the Houbara bustard nests in deserts and restricts turning behaviour to the cooler morning and evening parts of the day (see also Figure 5.5, p. 75) but time intervals between turning events are random within these periods. As with much about egg turning, it is unclear whether random turning events play any role in the physiological effects of egg movement and there is nothing to suggest that regular turning has any detrimental effect.

The recent research into the rate of egg turning in nests has highlighted how little we know about this behaviour. There is a need for further research into the angle and frequencies of turning for eggs of non-poultry species. The results of existing and new research man that turning control systems of many incubators may need to be re-designed to allow for different rates of egg turning.

Things to consider when buying an incubator

Incubators come in all shapes, sizes and costs. Although many incubators can be used successfully with a wide variety of species, your particular interest or application is going to dominate your choice. For example, breeders of exotic birds are seldom concerned about egg capacity because the number of small parrot eggs that will fit in a small table top incubator often exceeds the probable number available for incubation. So in this case, features such as automatic humidity control and a reputation for reliability and success may be much more important than size. A little realism about the investment compared with the potential value of the baby bird is helpful. No incubator can be guaranteed to hatch every egg but paying a little more for a better product can make good sense.

Breeders of parrots also have the consideration that the eggs from a single clutch are incubated by the parent as soon as laid, and hatch in sequence. So it is likely even if they are pulled and placed in an incubator, that eggs will be hatching at all sorts of different times. A small hatcher is very valuable here in keeping hatchlings away from younger eggs and from automatic turning systems, a benefit in hygiene and safety.

Sometimes how big the eggs are is important. Small table-top incubators will usually only hold eggs up to a certain size. Large waterfowl and ratite eggs may need a physically big incubator even though the egg numbers are small. If you are hoping to incubate eggs from a variety of species each laying different sized eggs then you either have a still-air machine for each egg size of you have a force-draught incubator.

If you want to control the weight loss of your eggs more closely then it is often a good idea to buy at least three identical machines. In this way you can set them up at the same temperature but at different humidities and move eggs around in order to optimise weight loss (see p. 147). The ability to adjust the rate of automatic egg turning may also be a consideration for people wishing to incubate eggs of species like parrots that produce altricial young.

For breeders of game birds or waterfowl the choice may be determined much more in terms of cost per egg capacity. These birds naturally have larger clutches and lay many more eggs in a breeding season. Larger table top models or cabinet incubators are likely to be suitable. Although cabinet incubators often have hatching trays at the bottom, consider investing in a separate hatcher. The capacity need only be about one third of the incubator capacity if eggs are being collected and set on a regular (weekly) basis. The benefits in hygiene and ease of cleaning are very significant.

Many breeders prefer forced-draught incubators to still-air machines; the control of temperature and ventilation is undoubtedly better and the availability of low price, polystyrene foam, still-air incubators has dented their reputation. However, still-air incubators can perform remarkably well if used correctly and offer an incubation environment rather closer to nature when set up well. To work well, all eggs need to be of approximately the same size.

A prime consideration is where you plan to keep your incubators. Some of the larger machines for ratite eggs are just too large to fit into some rooms. A large number of small incubators can take up a surprisingly large amount of space. You will also need to consider how you will provide fresh air for the rooms.

There are a variety of things to avoid. Cabinets which cannot be cleaned or disinfected easily, particularly those with long, inaccessible air paths can be a liability. The material from which the interior is constructed should be smooth and non-absorbent. Poor thermal insulation aggravates temperature differences around the egg chamber; the temperature might be rock steady on the thermometer but quite different in another place within the egg chamber. Check the type of fan(s) used; low voltage DC fan motors are virtually silent and vibration free and also produce very little heat. Badly balanced fans cause vibration, which is suspected of being an important cause of embryonic mortality. Egg turning systems are another potential source of vibration, but it is continual vibration rather than occasional disturbance to eggs, which is most likely to cause damage. Inflexible turning systems, which cannot easily accommodate various sized eggs, can be a nuisance in some cases.

An important thing to be sure of is that there is some independent check on the temperature control system. Dependence on a single sensor for control and monitoring is risky. If the sensor lies, you will not know until you have probably done serious damage.

Summary

- The main types of incubator are described.

- Construction and operation of different types of machines are discussed in relation to the cabinet, air exchange and movement, control of temperature and humidity, and turning.

- Emphasis is placed on maintenance of the correct measurement of temperature

- Factors that need to be considered when deciding about purchasing an incubator are described.

8 – Small-scale Artificial Incubation

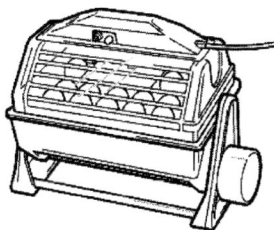

In this chapter I describe the basic procedures of artificial incubation on a small scale. Much of this information may seem obvious to people with some experience of artificial incubation but my intention is not only to describe procedures but also to explain the reasons why they are necessary and their importance in determining successful incubation. This information starts from the point that the egg is laid and follows through to hatching. It is hoped that this will provide invaluable information for the novice as well as better educating those people who already have experience of using small incubators.

Pre-incubation handing

Failure to hatch bird eggs in incubators will not always reflect the incubation environment those eggs had been exposed to. There are other factors before incubation has started that affect the quality of the egg and the viability of the embryo within it. How to recognise and assess many of these problems is discussed in Chapter 9. In general, the environment that the egg experiences before reaching the incubator is crucial to success or failure, irrespective of the incubation environment. The contrary is also true in that the best quality egg may fail to hatch because the incubation environment is incorrect.

Egg laying and handling

Before progressing how do I define a "quality" egg? In simple terms such an egg has no physical defects, is not lacking in any chemical component and contains a vigorous, viable embryo. A poor quality egg will most visibly have a defective shell and will have come from a female fed on an inappropriate diet although this is not obvious from outside appearance. Unfortunately, when the eggs are laid not all embryos are capable of developing through hatch. Here I will assume that the "quality" of the egg at lay is the best it possibly can be and that our actions can contribute to this loss of quality. Problems associated with loss of quality are discussed in Chapter 9 (see pp. 173–177).

Firstly, the environment that the egg encounters within the first few seconds post-laying is critical in whether it will remain free of microbial contamination (Box 8.1, p. 130). Research has shown that if a domestic fowl egg is caught as it is laid and is placed on a dirty nest environment for 30 seconds

BOX 8.1 – MICROBIAL CONTAMINATION OF HATCHING EGGS

Under ideal circumstances there are only a couple of times that bacteria, fungi or viruses can contaminate an egg. The first, and most important, is within a few seconds of laying. At this time the egg is wet and the pores are full of fluid. If the shell comes into contact with a dirty surface the film of water on its surface provides an ideal route for organisms to enter the pore canal and move down into the contents. The mode of contamination can be of either living organisms (mainly bacteria and viruses) or their spores (bacteria and fungi) that germinate once in ideal conditions. Good nest hygiene and prompt collection of eggs will help to minimise this risk. However, once the contents of an egg are contaminated then very little can be done to remove the problem. Cleaning and disinfecting the shell will sanitise the surface but will not affect the contents. Even fumigation with a gas, such as formaldehyde, that can traverse the eggshell will have minimal effects on contamination of the contents. However, leaving eggs dirty is not an option (see pp. 131).

During hatching is the second most likely time for an egg or the contents to be contaminated. Any dirty eggshells will pose a threat to newly hatched birds – they are wet and may have exposed yolks or navels that could be infected by organisms living on the shells of other eggs. Bacteria flourish in the warm humid environment of hatchers and the simplest way to minimise this risk is to prevent organisms entering the machines in the first instance. The primary sources of contamination are eggs and people. Eggs should be cleaned and disinfected soon after collection. Care has to be taken when handling eggs and chicks in different machines to prevent cross-contamination. It is imperative that hatchers are cleaned out thoroughly after each hatch.

Eggs should not be contaminated during the rest of incubation. The relatively dry environment of an incubator minimises microbial growth but where water is present then organisms can flourish. This is particularly so if there is organic material around to provide a source of "food". Dirt and droppings on the shell surface are ideal sources of "food" for organisms but in incubators it is usually broken eggs or material seeping from rotten eggs that provides the organic food source for micro-organisms. Only by keeping incubators clean will high standard of hygiene be maintained. Humidity trays or sprays should be cleaned and disinfected on a regular basis to prevent build-up of organisms and debris, like lime-scale, that can harbour organisms.

A keen sense of smell will assist in identifying any addled eggs before they start to ooze their contents or explode in the cabinet. Looking for bad eggs and removing them quickly is a good way to keep incubators clean.

Figure 8.1. Photograph of two halves of the inside of the same eggshell following incubation of the eggshell filled with a culture medium that stains bacterial growth. The half on the left was placed on a dirty nest box for 30 seconds immediately after being laid. The half on the right was placed on the same medium after the 30 seconds had elapsed and stayed there for 30 seconds. The dye indicating bacterial growth is clearly visible in the left hand shell. Photograph courtesy of Dr Nick Sparks.

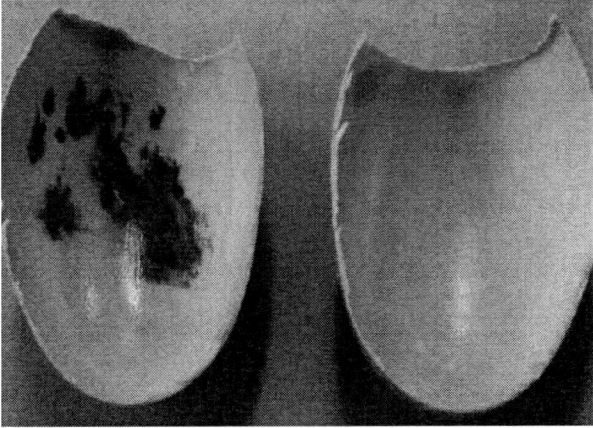

then the chances of it being contaminated are very high. If the same egg is turned over after 30 seconds and placed on the same surface for a further 30 seconds then the risk of contamination is very greatly reduced (Figure 8.1). This difference is caused by the fact that it takes the cuticle around 30 seconds to "cure" and become an effective barrier against bacterial contamination of the pores.

Ostrich eggs have no cuticle and my own work showed that ~40% of eggs set were lost to microbial contamination. The nest sites were on pasture in Britain – a very damp environment compared with the sandy nests of the ostrich's native Africa. By contrast, commercially produced pheasant eggs are commonly laid in very wet conditions and yet losses of eggs to bacterial spoilage are only around 1% of eggs set (but see p. 173). It is unclear why this is so but the cuticle on the shell must offer great protection against the movement of contaminated water into the pores. Grebe eggs take this to an extreme. They are incubated on floating nests of vegetation and so are partially submerged! Amazingly these eggs lose water during incubation and hatch well. Fowl eggs placed in the same nests go rotten within a couple of days. Nest hygiene is, therefore, very important when breeding birds in captivity.

It could be argued that natural nests are never clean and indeed they can have an extensive bacterial and fungal communities (Box 2.1, p. 15). The birds may not clean their nests but some species at least do attempt to keep microbial contamination down with the use of plant material with anti-bacterial properties (Box 2.1, p. 15). Furthermore, eggs regularly go rotten in nests! The risk of microbial contamination of eggs is ever present whatever the species and so it is always just good practise to have clean nest sites. This should minimise the risk and prevent any loss of egg quality within seconds of laying. Nests can get dirty very quickly and so regular monitoring and, if necessary, cleaning of nest sites and materials is important.

Good hygiene should also extend to the people picking up the eggs. There is no point in maintaining clean nests if the person picking up the egg comes along immediately after cleaning out bird cages and hasn't followed the simple hygiene practise of washing their hands. The highest hygiene should be maintained around eggs to prevent contamination of clean, newly-laid eggs as well as those already collected. There are many chemical products aimed directly at hand hygiene. Developed for hatcheries in the poultry industry, these antimicrobial sanitising solutions, soaps and hand rinses are worth using. However, soap and water will do if there is nothing else.

What if the egg is dirty to start with? Here there is a range of things to do. The first is do nothing – the egg contents may already be contaminated and trying to disinfect it is pointless. Whilst, this is a valid argument, eggs covered with droppings, soil or nest litter pose a real threat to all of the other eggs that they could come into contact with during incubation and hatching. Moreover, even if the egg contents are not contaminated but the dirt could contaminate the chick as it hatches or other chicks in the same machine. Doing nothing is not really an option.

Why not fumigate the soiled eggs? Here the eggs are placed in an environment rich in a gaseous fumigant, *e.g.* formaldehyde, ozone or hydrogen peroxide, or in a fog of a sanitising solution (Box 8.2, p. 133). The latter is created by a spray system that provides a very fine mist of the sanitising solution that is able to diffuse easily around all surfaces of eggs (and incubators and rooms). The problem is that any disinfectant or sanitising solution used on eggs is only fully effective if the surfaces are already clean. All of these chemicals kill on contact (Box 8.2, p. 133) but they will not destroy any organisms they cannot come into contact with. Therefore, the surface of the shell under a blob of mud or droppings will be unaffected by any gaseous disinfectant. Any organisms that have penetrated the pores and into the egg contents may be killed by any gas diffusing into the pores but will not be affected by fogging solutions because they tend to condense on the shell.

An alternative is that the offending material could be rubbed off and the egg could be left or the egg could be fumigated with a gaseous sanitising chemical. This may remove the source of the problem but does not fully clean

BOX 8.2 – TYPES AND MODE OF ACTION OF SANITISING CHEMICALS

First a few definitions. "Hygiene" is the science of maintenance of health by cleanliness. A "disinfectant" is a chemical designed to kill micro-organisms whereas a "detergent" is a chemical designed as a cleansing agent. A "sanitising" chemical is designed to promote hygiene and hence has both detergent and disinfecting properties. There is a wide range of chemicals on the market for use in the hatchery industry and they vary in their chemical composition and properties. I am unable (and unwilling) to provide a comparison of different products on the market but rather here I will briefly describe the main types of chemicals in use and how they are of benefit.

Formaldehyde is a strongly reactive gas that for many years has been commonly used to fumigate eggs and machines. It has highly effective disinfectant properties but no detergent property and so works best on surfaces that have already been cleaned. However, the noxious smell and high toxicity of formaldehyde has made its use less popular in recent years. As a result other highly reactive but less noxious molecules, such as ozone and hydrogen peroxide, are being investigated as replacements.

Chlorine and hypochlorite based chemicals are commonly used as disinfectants but they are not very effective against fungi or viruses and they have poor detergency. They are also highly corrosive. Sanitising compounds based around quaternary ammonium chemicals are only effective against bacteria and have low toxicity. However, they have a good detergency and are not corrosive but are relatively expensive to manufacture. Compounds based on phenolics are also common and are effective against bacteria and fungi. These chemicals are highly toxic and have low detergency. One alternative method of disinfecting eggs and dry surfaces is exposure to light intensity ultraviolet (UV) light but exposure is also harmful to man and is only effective on contact. Any organisms in any shadows will be unaffected.

All of these chemicals (and UV light) kill micro-organisms by attacking and destroying the structure of organic material (*i.e.* chemicals based on carbon). This is why they kill organisms but once a chemical molecule has reacted then it is spent and cannot react again. Unfortunately, disinfectant molecules will also attack non-living organic material and can be de-activated by inert material in faeces and soil. It is for this reason that it is important to clean away faeces and soil before or at the same time as application of the disinfecting compound (see pp. 133–134).

One thing is certain the ideal sanitising product for a hatchery (of whatever size) has yet to be produced. Disinfectants should be used in conjunction with detergents to maintain a high standard of hygiene at all times. It is only by doing this will the hatchery environment remain free of large numbers of micro-organisms.

or sanitise the eggshell surface and kill any organisms present. Rubbing away dirt will not fully clean the surface. "Buffing" is a common practise – here the shell surface is rubbed "clean" with an abrasive pad or sandpaper. Again the process does not clean well nor does it sanitise the surface. More importantly the abrasion creates fine dust that can be pushed down into and block the pore openings. This can cause problems with respiratory gas exchange towards the end of incubation (see pp. 169–171).

The ideal situation is to wash the eggs clean of any offending material and then sanitise them. Egg washing is very controversial but many of the adverse results can be usually attributed to people not carrying out the procedure properly. Poor practise can give egg washing a bad name because it can and will cause more problems that it was designed to solve. The ideal way of washing eggs is as follows.

The sanitising chemical should be pre-dissolved or mixed and the washing solution should be at 40°C before proceeding. At this temperature the solution is always going to be warmer than the eggs. Therefore, as the egg is washed the air in the pore canals warms and expands outwards forming bubbles that can be easily seen on the shell surface, effectively helping to prevent any solution from entering the pores. Eggs should be washed carefully to remove any adherent material and then removed and rinsed in a second solution of the sanitiser, again at 40°C, before being placed on a clean surface to dry. Ideally, it is best to use an egg washer that has a thermostatic control of the solution temperature. Excess solution can be removed with clean disposable paper tissue or the shells can be carefully dried with a hair-drier set on the lowest heat. Thereafter, the eggs should be only handled with clean hands. Many people wear rubber gloves but these can be slippery when wet and can easily get contaminated. I much prefer using freshly washed hands to handle eggs. This seems a very simple procedure – whatever could go wrong?

The list of potential problems is rather long but here are a few things to look out for. The egg goes bad because its shell was cracked before it was washed or was cracked during washing. The washing solution is not at the correct temperature. Too warm (*i.e.* over 50°C) and immersion for too long increases the risk that the embryo is damaged by the heat. Too cold (*i.e.* below the temperature of the egg) and the risk is that the air in the pores will contract thereby pulling in some of the washing solution. In this case the solution itself may damage the embryo or the solution may be defunct and bacteria are drawn into the pores.

A common problem is the habit of washing all eggs in the same solution. Ideally you should wash the cleanest eggs first. Any really dirty eggs should be left until later so that they do not soak up the entire disinfectant chemical and render the washing solution useless. Here is an even more radical idea to consider. What about disposing of exceptionally dirty eggs (my criterion for these is whether I would consider eating such a soiled egg) on the basis that the risk of contamination is already very high? The remaining eggs will be sanitised more effectively and the risk of contamination of other eggs from eggs going bad during incubation will have been reduced.

The strength of some chlorine-based chemical solutions can be tested during washing but it is better to change the solution on a regular basis rather than trying to improve the disinfectant properties of an already dirty solution. Leaving eggs in the chemical solution for too little or too long can cause problems, particularly in conjunction with changes in temperature. Leaving a small egg in a solution at 40°C will actually warm the egg contents considerably (see pp. 85–86) and this heat will be retained in the egg for some while after it is removed from the washing solution. These conditions may raise embryo temperature to such a level that it starts development but not at a rate that can be sustained long enough for normal development. If you leave eggs in the washing systems that are not controlled by a thermostat, then the solution will cool and the danger is that the eggshell will get saturated with water (shells often develop translucent spots when viewed with a candling lamp).

"Hatchery" design and operation

Whilst considering hygiene of eggs it is useful to turn to the "hatchery" where egg handling and incubation will take place. Modern commercial poultry hatcheries are huge buildings very much designed with hygiene in mind. There is a definite flow of eggs through the building from a reception area, into a store before going to the incubators and then the hatchers. Chicks are then taken out of the hatchery on the other side of the building from the egg reception area. In this way there is minimal chance of cross-contamination of clean eggs by the fluff from hatched chicks.

These principles apply equally to small-scale operations. If at all possible it is good to have a cool room to store eggs, with a different room for the incubators and a separate room for the hatchers. Ideally you have a separate hatcher rather than using the incubator (unless the eggs are incubated single stage). This will reduce the chances of cross-contamination of clean eggs with hatch debris. If at all possible these rooms should be dedicated to incubation purposes and through traffic of people should be avoided. It may not be possible but it is best to avoid the situation where you have to go through the egg store and incubator room to reach the hatcher room. Foot baths of disinfectant are a useful way of preventing the spread of organisms from bird pens to the hatchery.

These rooms may need to be air-conditioned to ensure that the correct environment is provided for the machines and there should be a supply of fresh air at all times. Having a window open may not be sufficient to supply the amount of air and a small window-mounted fan may be necessary. This will also help circulate air around the room. For one machine in a large room lack of room aeration is not a great problem but a room dedicated to incubators should have good aeration to prevent the air from getting stale. Air-conditioning will also help to control ambient humidity although if you are incubating eggs from desert-nesting species then the room air may need to be de-humidified.

Hygiene of eggs and machines is very important in prevention of problems with micro-organisms. If eggs are accidentally broken then the debris should be cleared up quickly and any contaminated surfaces sanitised. The people working with eggs are a source of contamination and there should be minimal contact between breeding birds and the incubators. Personal hygiene of people handling egg will prevent contamination of eggs but will also stop eggs from infecting people. Considering where eggs actually come from, people should always wash their hands thoroughly before and after handling eggs.

Storage

Bird eggs are naturally built to withstand a period of time before incubation starts without any loss of embryonic viability. This allows the female to build up a clutch of eggs before it has to start full-time incubation. As a consequence, we can also store eggs in order to ensure that we have sufficient numbers for incubation. This may be important for precocial species of bird that can be more easily reared in-groups. On the other hand, most parrots and other species commence incubation soon after the first egg is laid and each egg in the clutch will be at a different stage and hatch at a different time.

There may also be another advantage to storing eggs for a couple of days after laying. In the commercial domestic fowl, eggs have a better hatchability after a storage period of 3 days than when they are set immediately after laying. In the fresh egg the thick albumen capsule (see pp. 25–26) is an effective barrier to diffusion of chemicals within the egg contents and in particular oxygen diffuses very slowly though this layer. This lack of oxygen may restrict early embryonic development and increase the chances that weak embryos will die. It is not known whether storage of eggs from more exotic species has the same effect. Indeed a few days storage may benefit albumen-rich eggs of semi-altricial and altricial species. However, if eggs are kept under good conditions for only 2–3 days there will not be any loss of embryonic viability.

Good storage conditions require a stable temperature of between 15–18°C and a relative humidity of around 75%. The low temperature prevents any cell division in the embryo and the high humidity helps minimise loss of water vapour during the storage period. Under these conditions poultry eggs will remain viable for 7 days and there is no reason to suggest that eggs from other species will not do the same. Storage for longer periods tends to reduce embryonic viability but this can be minimised by storage at a lower temperature (12°C).

Storage at temperatures above 21°C will lead to partial development and perhaps take the embryo beyond the gastrula stage of early development (see p. 47) and reducing its long term viability unless normal incubation temperatures are adopted. Holding eggs for several days at temperatures of 25°C or above is to be avoided at all costs. Storing eggs at low temperatures is not a good idea either. Embryos have a poor tolerance of temperatures below 10°C and as temperature approach freezing there can be damage to the embryo or to the integrity of the yolk and albumen. Below freezing formation of ice crystals in cells kills the embryo. There are some people who use a domestic refrigerator to store eggs but the temperatures maintained inside (typically at 4–5°C) would not generally do eggs too much good over a long period of time. A couple of days of storage in a refrigerator may be acceptable if there is no other place to keep eggs. It would be much better to store in a cool room or cellar.

Other factors improving embryonic viability during storage involve covering trays of eggs with plastic bags (to raise humidity), keeping them in plastic bags flushed with nitrogen (to remove oxygen from the air), or setting the eggs upside down (to prevent the embryo drying out). Another technique is to put the stored eggs into an atmosphere rich in carbon dioxide for the 24 hours before setting. Although shown to have benefits in commercial-scale operations, these methods are rarely used because they are time-consuming and relatively expensive. The value of the eggs involved does not often justify using the techniques. However, the value of some of the eggs incubated in small-scale operations may merit the use of these techniques to maximise embryonic viability during prolonged storage.

Keeping eggs leads to a variety of changes within the egg that may endanger the embryo if storage is prolonged. Firstly, there is a loss of carbon dioxide (CO_2) from the egg. There is a lot of CO_2 dissolved in the body tissues of birds and this gets incorporated into the egg contents. Once outside the body the high CO_2 levels inside the shell contrast sharply with the very low levels of CO_2 in the air (0.03%). This leads to a rapid loss of CO_2 from the egg contents, which reduces the acidity of the albumen and yolk – CO_2 dissolved in water forms weak carbonic acid. The loss of acidity is so great that the albumen becomes highly alkaline (more than pH 9.0) and this may compromise the viability of the embryo's cells. Holding stored eggs in high levels of CO_2

allows this gas to re-enter the egg and the less alkaline albumen may assist in promoting embryonic development.

Prolonged storage also causes the breakdown of the thick albumen capsule, which then allows the yolk to float nearer to the inner shell membrane at the top of the egg. There is then a danger that there will be mechanical damage to the embryo by physical contact with the membrane. Furthermore, the thinner layer of albumen above the embryo is more prone to loss of water vapour through the eggshell and there may be dehydration of the albumen and the embryo. Holding eggs with the sharp pole upwards traps a relatively thick layer of albumen over the embryo and may prevent the embryo coming into contact with the inner shell membrane. Problems of dehydration, exposure to oxygen or mechanical damage will then be reduced.

Finally, closer proximity of the embryo to the inner shell membrane increases the risk of the embryo coming into contact with gaseous oxygen. Although essential for life, oxygen is a highly reactive molecule, which can split into two oxygen atoms. These oxygen "free radicals" act like bullets cutting through tissues and cells causing great damage.

The exact reason for the loss of viability of eggs under long periods of storage (*i.e.* more than 7 days) may be one of these reasons or a combination of them all. On a small scale the effects of storage will be minimised by maintaining the correct temperature and humidity. Furthermore, holding eggs with the sharp pole upwards would be a useful procedure.

Turning of eggs perhaps once or twice a day is another useful practise during long term storage. Here the embryo is regularly moved within the egg and does not have to suffer any particular environment for any length of time. This may assist in maintaining the appropriate environment around the embryo so that it can start embryonic development straight away once the correct incubation temperature is reached.

An important aspect of egg storage is the pre-warming of eggs prior to setting into the incubator. Cold eggs will take longer to bring up to incubator temperature (although this will depend on egg size) and there can be problems with condensation on the shell surface. Putting a cold egg into a warm humid incubator will lower the temperature of the air immediately around the egg and the air's humidity will rise. Should this temperature drop be large enough it will cross the dewpoint temperature (*i.e.* the temperature at which the air is saturated with water vapour − 100%RH) and water vapour will condense out of the air forming dew on the eggshell surface. This layer of water can allow any micro-organisms on the shell to grow and move around. The film of water extends into the pore canal and provides an ideal route for motile bacteria (*i.e.* those that swim) to cross the shell and infect the egg contents.

This situation is resolved by pre-warming the eggs at room temperature for a few hours before setting. Holding the eggs at 22–24°C will raise their temperature to a point above the dew-point and once in the incubator the eggs will not cause any condensation. The eggs will also warm up quicker. An egg the size of a partridge (~20 g) will take around 4 hours to warm up compared with 8 hours for a fowl egg (~60 g) and 24 hours for an ostrich egg (~1,500 g).

Collecting eggs from birds

Many eggs are removed from the nest within hours of being laid and are either stored or set to incubate. In this situation the egg is able cope with either option. The typical response of the female is to produce more eggs and this "egg-pulling" is a good way of increasing egg production over a laying season.

However, some eggs are left, either accidentally or on purpose, under the parent birds for a period of incubation. In this case, any eggs that are pulled have to be set to incubate relatively quickly. Letting the egg cool down for a few minutes is not a problem to the embryo. It is not unusual for eggs to cool when the parent leaves the nest. However, the egg has to be returned to an incubation temperature as soon as possible. You cannot store eggs that are partially incubated.

Some people may wish to take eggs way from a poor parent but return the eggs to the bird around the time of hatching. For instance, the parents are much better at rearing the chicks even in captivity. In this circumstance the birds cannot be left without any eggs to incubate. Removal of eggs without providing a replacement is the same as the eggs being eaten by a predator. The visual and physical stimulation of the eggs will be lost and circulating levels of prolactin (see p. 67) will be lost and the bird will probably return to breeding condition in preparation for producing another clutch. Trying to put eggs back in the nest later just won't work. Only by providing incubating birds with dummy or infertile eggs will broody behaviour be maintained for long enough to return eggs for hatching. Luckily incubation behaviour does not rely on having living eggs under the bird.

There is one more important point to make about pre-incubation conditions before going to describe incubation. Eggs are incredible structures being strong enough for the birds to sit upon when incubating yet weak enough for a chick to break out of. Outward appearances are deceptive but bird eggs are living organisms and need to be treated with respect at all times. There is some evidence that shows that birds are far from gentle with their eggs and yet the embryos seem to be able survive this. This physical "abuse" of eggs is rarely sustained or that great. Problems can arise prior to incubation with egg-shells being cracked by rough handling. Shaking of eggs can disrupt the contents and kill the embryo. By contrast, during incubation the odd jolt experienced by an egg as it turns is very rarely anything to worry about.

I like to think of eggs as being like bone china. This is strong enough to eat from but it always needs to be handled very carefully to prevent chipping or any other damage. By being careful you will not destroy the quality of the egg before or during incubation. Even if the birds are rough with eggs this doesn't mean that we have to be.

Creating the incubation environment

When are eggs are set to incubate, be it in a multi-stage or single stage system (Box 6.1, p. 102), the operator of the machine has to ensure that the correct incubation environment is provided. Ideally, incubators should already be warm when the eggs are set because this assists in raising the eggs' temperature as quickly as possible. It is a good idea to intersperse eggs of different ages within multi-stage incubators so as to help produce a uniform temperature distribution around the machine.

Temperature

The accepted incubation temperature for bird eggs is typically between 37.0–38.0°C with a normal working average of 37.5°C. There are many reports of the temperature recorded for the middle of eggs in nests. The range is 31.6–39.8°C and average 35.7°C. However, these temperatures invariably do not represent embryo temperature and reflect variations in fertility, age of the embryo and the size of the egg.

Development will only proceed only within the narrow range of temperatures (37.0–38.0°C) for most bird species but the control system of an incubator needs to ensure that it is registering this as the egg temperature. In still-air incubators it is critical that the tip of the temperature sensor lies at the top of the egg (not touching) and the temperature is set slightly higher than in force-draught incubators, often 39°C. The manufacturer's recommendation should be checked because machines have different temperature gradient characteristics. In a force-draught incubator then the sensor has to be placed in the air stream to ensure that accurately records the temperature moving over the eggs (see pp. 120–122).

The main exceptions to an incubation temperature between 37.0–38.0°C are eggs of the ostrich, emu, rheas, cassowaries and kiwis. Artificial incubation of these large eggs is only really successful at a lower set-point temperature typically between 36.0–36.5°C. This does not reflect a difference in the actual temperature for embryonic development between ratites and other birds but rather is a consequence of incubation of these larger egg sizes in artificial incubators. The problems of trying to incubate an emu egg at 37.5°C do not lie at the start of incubation but at the end. The large egg volume retains heat very effectively partly due to the relatively small surface area for heat exchange. At a temperature considered normal for other birds the emu egg retains heat once the phase of embryonic growth starts and egg temperature is raised well above the set point. Incubating the eggs at a lower temperature means that the machine continually acts to cool and remove the heat energy. In table-top incubators the eggs would heat the air rather than the heaters and so would be cooled in the process. Under a single stage programme the initial temperature can be within the range typical for other birds but as development proceeds the set point temperature has to be dropped. I have set ostrich eggs under single stage incubation with a starting temperature of 37.0°C without any ill-effect. Large eggs of other bird species, *e.g.* geese, swans and larger penguins, will also need a lower set-point temperature under artificial incubation.

Getting the incubator temperature right is so critical because of the embryo's need to be kept with a narrow range of temperatures. Although birds are warm-blooded (*i.e.* they can generate sufficient heat in their own bodies to regulate their body temperature well above ambient), bird embryos are cold-blooded with no ability to change their body temperature as the ambient con-

Figure 8.2. Effect of reduced air temperature on the heart rate of phesant embryos on day 20 of incubation. Times at which set-point temperatures were changed by 5°C are shown by the arrows.

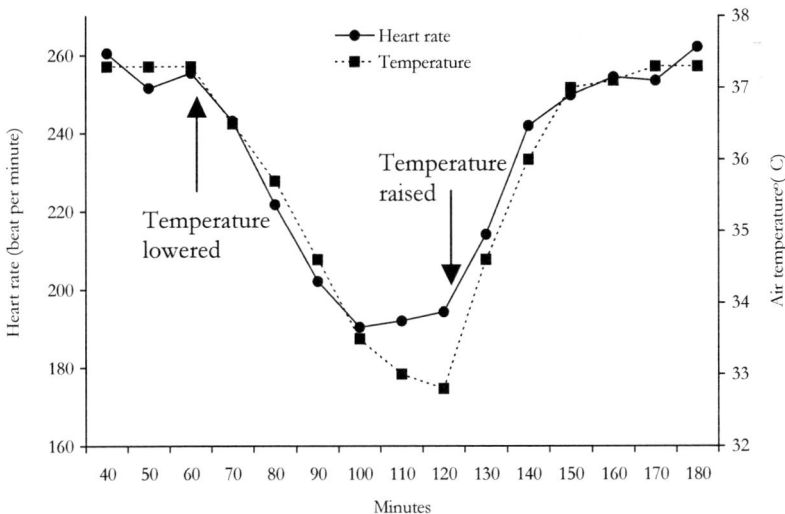

Figure 8.3. Correlation between incubator air temperature and heart rate of phesant embryos on day 20 of incubation. The line shows the trend.

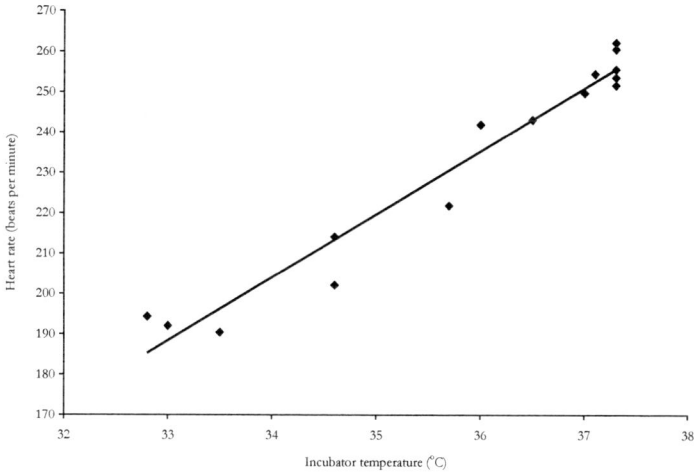

ditions change. Cool down a bird egg and the embryo's metabolism will slow down, warm the egg and the embryo will do things faster (within reason). This point is well illustrated by the effect of reduced incubator air temperature on the heart rate of pheasant embryos (Figure 8.2). As temperature of the air drops there is a rapid reduction in heart rate that stabilises at the new stable temperature but rising rapidly once the normal set point is restored (Figure 8.2). There is a good correlation between heart rate and air temperature and for in pheasants each 1°C drop in temperature there is a drop in heart rate of around 15 beats per minute (Figure 8.3).

Only during the last few days of incubation have periods of cooling been shown to stimulate a very limited thermoregulatory response in some embryos. Hiroshi Tazawa and colleagues have shown that as the egg cools precocial duck and fowl embryos can generate a small amount of additional heat in an attempt to raise body temperature and as such this mimics the response of a free-living neonate. However, in a semi-precocial species, like the brown noddy, then this response is extremely small and in altricial pigeon embryos there is no heat generated in response to a drop in egg temperature. Indeed, true thermoregulation is not possible in pigeons until several days post-hatching.

Therefore, for most of the time embryos in incubators have no choice in this response to temperature. At temperatures cooler than the optimum the rate of embryonic development is slowed and the incubation period is extended (Figure 8.4). Moderate increases in incubator temperature accelerate

Figure 8.4 Relationship between incubation temperature and the length of the incubation period of domestic fowl eggs.

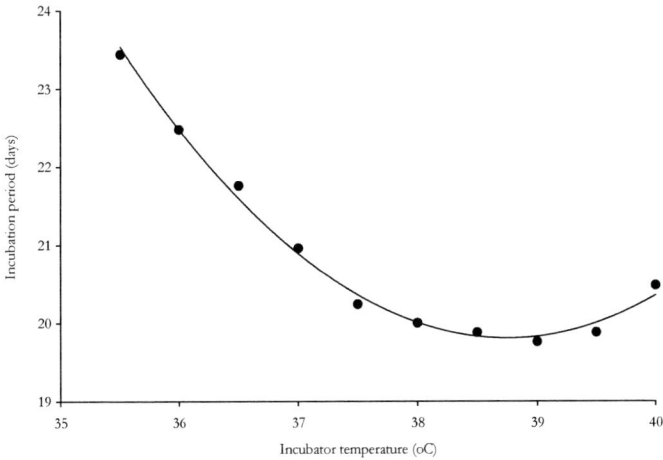

Figure 8.5. Effect of incubation temperature on growth in fowl embryos up to the day before hatching.

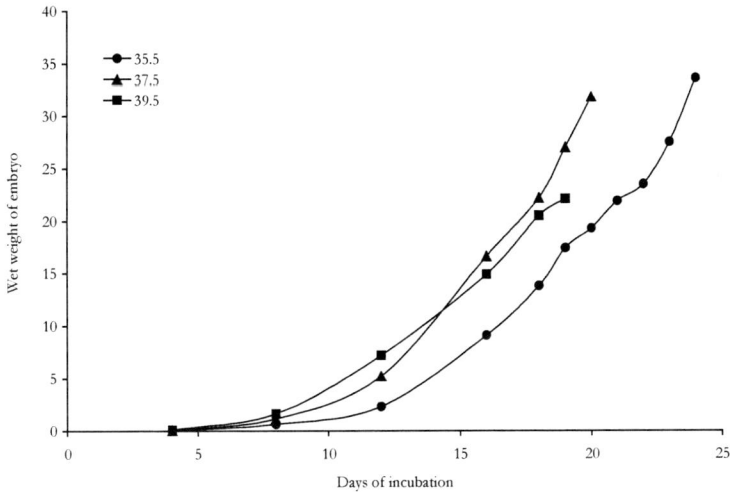

the metabolic rate and shorten the incubation period (Figure 8.4). However, as the temperature continues to increase then the embryo's rate of development becomes adversely affected and the incubation begins to get longer (Figure 8.4). There is also a massive decrease in hatchability. Although this has only been documented in domestic fowl embryos, it is also seen in reptilian embryos, which are more tolerant of different incubation temperatures, and so it is reasonable to assume that this pattern will apply to bird embryos in general.

This pattern is also observed for embryonic growth rates (Figure 8.5). At 37.5°C the growth rate of fowl embryos is normal but a two degree drop in temperature slows the rate of growth and those birds that survive to hatch emerge 4 days later than normal. A rise in set point of two degrees shows an initial acceleration in growth rate but this falters during the second half of incubation. Those few chicks that survive are small when they hatch a day before the birds incubated at the optimal temperature (Figure 8.5).

These responses highlight the sensitivity of embryos to higher than optimum incubation temperatures. Artificial incubation has to strive to minimise this problem during incubation by getting the balance between heating and cooling. In small table-top incubators heating is from the heaters and the eggs themselves generating metabolic heat. By contrast, cooling is achieved by loss of heat through the cabinet walls as well as loss of warm air from the machine being replaced by cool fresh air and absorption of heat by cooler eggs that a less advanced stage. It is critical that the temperature sensor actually measures the temperature that is representative of the eggs and the difference between egg temperature and sensor temperature is minimised.

Egg temperature in many species is not constant during natural incubation because the incubating parent periodically leaves the nest and exposes the eggs, which then cool (see pp. 71–78). Whether this periodic cooling has any significant role remains untested but several people are known to try to mimic the cooling period under artificial incubator. This can be removing the lid from the cabinet, opening cabinet doors or by having the heater or machine on a time switch so that the power is periodically cut. The degree to which these procedures mimic egg cooling in a nest remains unclear. It is possible that periodic cooling, and perhaps temperature gradients in eggs, have some important role in proper embryonic development in some species. Poultry species may not have such a requirement or perhaps modern strains have lost this sensitivity. I know that mallard eggs apparently require 30 minute periods of cooling from 14 day to day 21 of incubation for optimal hatching. However, it is not known whether this cooling or the spraying of the eggs with water just before returning them to the machine, is critical.

However, the period of cooling in a small song bird nest may not be as simple as is first imagined. It can be clearly seen in Figure 5.2 (p. 72) that in some instances egg temperature rises before the bird leaves the nest (see arrow in Figure 5.2). Therefore, the female appears to be investing the eggs with extra heat energy before it leaves the nest, perhaps to minimise the effects of cooling. Furthermore, as is shown in Figure 5.10 (p. 72), small altricial eggs are generally much warmer than larger precocial eggs whilst the bird is present and thereafter. Finally, my own unpublished observations show that attentiveness in songbirds does not correlate with the length of the incubation period – embryos in eggs incubated for 75% or 100% of the time take the same time to develop through to hatch suggesting that cooling in these small eggs has little effect.

Such uncertainties about the role of cooling make it difficult to recommend that eggs are cooled periodically or that a temperature gradient is critical for normal development. However, short period of cooling will have little effect on the development of most avian embryos and so periodic cooling will probably do little harm. More scientific study of these aspects of incubation are necessary before I can say that periodic cooling is beneficial and the development of the contact incubator (see Chapter 10) will make testing these ideas easier.

Humidity

Provision of humidity in small scale incubators usually depends on two methods. Either water pans are placed within the cabinet, which provides relatively little control of humidity conditions, or water is pumped into the cabinet upon demand, which gives more controlled and sustained humidity. Whatever system is used control of humidity is an active process that should be checked on a daily basis. Furthermore, it is very important to clean and disinfect humidity pans and spray nozzles on a regular basis. Both systems are much more effective if there is monitoring of weight loss during incubation and assessment of whether humidity within the machine is within acceptable limits.

Monitoring weight loss is an important aspect of artificial incubation and yet many people appear to be unable or unwilling to find the time carry out this simple procedure. However, they ignore it at their eggs' peril. Getting the humidity wrong can severely reduce the chances of individual eggs hatching (see p. 167). Given that most people using small scale incubators are dealing with relatively few eggs it is curious why monitoring weight loss is considered to be of little significance.

Whether an egg hatches or not depends on a variety of factors (Chapter 9) with weight loss being very important. For any egg there is a high probability of it hatching if its weight loss is between 10–20% of initial egg mass (Figure 8.6). This makes sense for an individual embryo because it does not want to be a position where it has to lose an exact weight loss or its dies. Even though a weight loss of 15% from laying to pipping could be considered as ideal for an individual embryo it doesn't much matter if the weight loss is 12% or 19%. In nature, chicks have hatched from eggs with weight losses in the range of 5–

Table 8.1. Influence of hatchling maturity on average % weight loss.

	Average % weight loss	SD (%)
Precocial	14.3	2.0
Semi-precocial	15.4	2.3
Semi-altricial	14.9	2.8
Altricial	15.3	2.9

33% but the average values are remarkably similar between species (see Table 5.3, p. 88). Indeed, the average values for average % weight loss for different hatchling maturities are all very close to 15% (Table 8.1). However, at a weight loss below 10%, and above 20%, the chances of the embryo hatching decrease considerably (Figure 8.6). The reasons for this reduction in the chances of hatching are different for each end of the spectrum.

If eggs lose too much water during incubation indicates either the humidity in the machine was too low or that the eggshell had a high porosity (water vapour conductance; see pp. 33–35). In these shells the embryos face the problem of dehydration of the contents or their tissues. That is not to say that these eggs won't hatch but it all depends on them developing within a machine set at the appropriate humidity. Cracked eggs are the most obvious cause of dehydrated eggs and so it is critical that they are not set to incubate. Many people are keen to mend cracks in the eggshells of valuable species in the hope that the eggs will hatch. This is fine but only if the nest site was clean and the contents have not been contaminated by micro-organisms. Repairing such shells will only cause problems with the contents going rotten or the embryo developing yolk sac infection at the end of incubation (see pp. 173–174). Even if the shell is repaired the egg will still require regular weighing to ensure that the humidity in the incubator is correct.

Insufficient weight loss indicates either the humidity was far too high or the eggshell had a low porosity. Problems with low porosity eggshells are not primarily due to insufficient loss of water vapour but rather reflect problems with the uptake of oxygen by the egg (see pp. 169–171). Maintaining a low humidity may restore the correct rate of weight loss for these eggs but it will not affect the diffusion of oxygen across the shell and the embryos may still fail to develop normally.

One way of assessing weight loss is by looking at the size of the air space by candling the egg. The bigger the air space the more water the egg has lost. Making this subjective assessment of relative size is certainly a skill and I admit that it is not one that I have mastered in over 20 years of working with eggs. I much prefer to remove any guesswork by weighing the eggs.

Figure 8.6. Relationship between percentage weight loss during incubation and percentage hatchability.

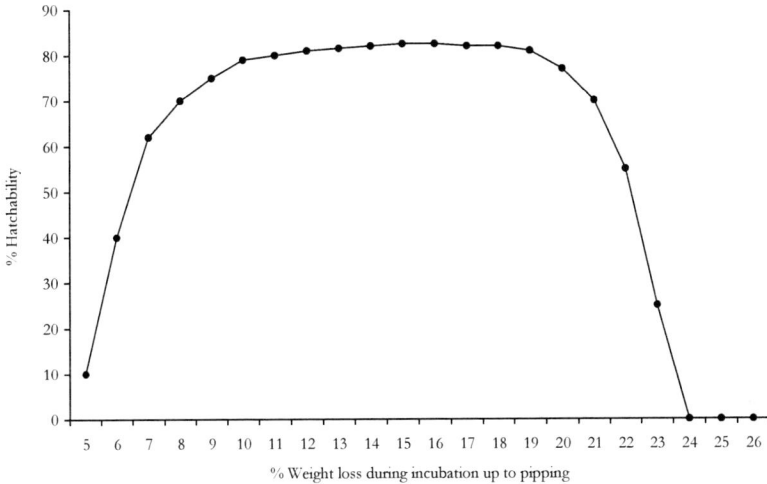

It is true to say that the most accurate way to tell whether the humidity employed in your machines is correct for the eggs is by weighing them. All eggs will vary around an average and female birds will respond to the climate of their local environment when forming their eggs. Hence, grebe eggs are incubated in wet nests and so have very high porosity shells that allow normal rates of weight loss despite the extremely high natural humidity. The reverse is true; desert-nesting species have eggshells that have relatively low porosity values that prevent excessive rates of weight loss in the dry nesting environment. Domestic fowl hens kept at sea level will lay eggs with shells with porosity values typical for the region. However, if the hens are transported 3,000 metres up a mountain then they change their eggshell structure in order to increase porosity to allow in more oxygen when incubated in this more rarefied atmosphere. On a smaller scale it does not follow that the humidity used in an incubator by the coast will suit eggs laid further inland.

Beware of people stating that the humidity is wrong for your eggs unless they can demonstrate that the weight loss is incorrect. All eggs need to have a starting humidity but this should be modified in light of actual rates of weight loss from eggs under incubation. Although it may seem time-consuming using the eggs to monitor humidity is the most accurate way of achieving the correct rate of weight loss.

Monitoring weight loss

It is important to realise that within an incubator the humidity provided has to provide the best environment for the *average* egg. In any group of eggs there will always be a few at the extremes of the range of variation in porosity but most will be close to the average value. The relatively wide range of weight losses acceptable to an individual embryo means that, at an average weight loss of 15% to pipping, most of the eggs will lie within the 10–20% values and will not be adversely affected by the average humidity. Those eggs at the extremes of shell porosity will suffer anyhow but they are usually small in number. This does mean that you have to monitor weight loss of either all of the eggs or at least a representative sample of the eggs within the machine. Taking one or two eggs means that you run the risk of selecting unrepresentative eggs that will skew the humidity conditions away from those required for the bulk of the eggs.

Monitoring egg weight loss is easy. You need a good set of scales (electronic ones are easier and quicker to use), the accuracy of which depends on the size of your eggs. Accuracy to 1 g is fine for a 1,500 g ostrich egg but if your eggs are only 10 g then you need a set of scales accurate to 0.1 g.

Immediately before setting the eggs you weigh your selected group and calculate the average egg weight. These eggs can be weighed individually if you wish to check weight loss per egg, or as a whole group (in which case a balance accurate to 1 g is often okay) and an average egg mass is calculated. The eggs are then set in an incubator at a given humidity. Most eggs will be fine at 50%RH in the first instance *so long as the weight loss is recorded a few days after*. Species that lay in relatively humidity conditions may need higher initial relative humidity of 60% and those in dry conditions may need an initial humidity of 40%RH. The eggs are left to incubate and after 3–4 days, or perhaps 7 days if the incubation period is long, the same eggs are weighed and a new average calculated.

Understanding whether the eggs have lost the correct weight loss depends on working out the daily rate of weight loss up to the day of external pipping (*i.e.* when the shell is broken). Therefore, let us say that I am incubating eggs of a species that average 20 g at the start of incubation and have an incubation period of 18 days. If you don't know when pipping occurs then it is easiest to assume that it will happen a day before hatching. My eggs should lose 15% of their initial weight, *i.e.* 3 g (20 g x 15/100) in 17 days. This is 0.176 g per day or 0.71 g every 4 days and most easily visualised as a graph (Figure 8.7). The eggs are then weighed after 4 days of incubation and the average weight is 18.95 g. This would produce a %WL of over 21% if left unchecked and indicates that the starting humidity is too low. Therefore the humidity is increased (Figure 8.7). At 12 days of incubation the predicted weight loss is now only

Figure 8.7. Lines indicating the pattern of percentage different weight loss values for a 20 g egg. The dashed line indicates the weight recording for an egg during incubation and the arrows show where the humidity was changed.

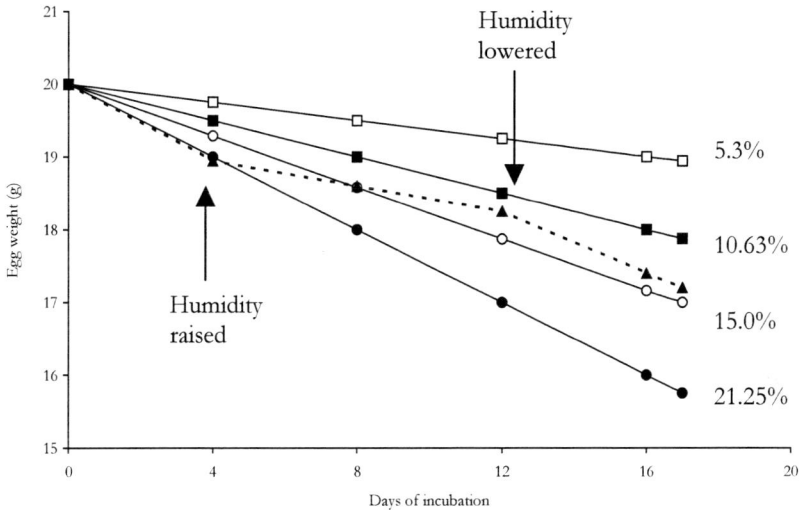

10.6% and so the humidity has to be lowered again to bring the actual weight loss at pipping on day 17 near to 15% (Figure 8.7).

The adjustments to humidity are usually not large. Most bird eggs will incubate within the range of 40–60%RH and so a change in RH of only 5% will have significant effects on the rates of weight loss. The further away from the line for predicted weight loss at 15% the actual values are, then the bigger the change in humidity has to be. This system works well if the eggs in the incubator are all about the same age and the same species but if there is a mixture of ages and types of eggs in the same cabinet then having one humidity will almost certainly not suit all of the eggs all of the time.

The solution to this problem is to have more than one incubator (of course this depends on whether you can afford the machines). Ideally, three identical machines are used, each set at the same temperature but maintained at different humidities. Machine "A" has a low humidity, "B" has the average humidity, and "C" has a high humidity. In my example above the egg would start in incubator "B" before being moved to incubator "C" at four days so as to benefit from the higher humidity. At 12 days the egg would be returned to incubator "B". In this way weight loss can be manipulated for individual eggs.

Does all this time and effort make that much difference? That depends on how much you value the chicks that your eggs produce. For batches of eggs, the further the average weight loss is from the optimum of 15% then the more eggs will be exposed to humidities that are not good for them. Higher than average weight losses tend to dehydrate eggs where low rates of weight loss cause water retention by the chicks that in extreme cases can prevent the chick from hatching. Incorrect weight loss not only affects hatchability but also chick quality. If weight loss is too little, because of high humidity rather than low eggshell porosity, then the egg has additional water that the embryo has to deal with by storing it in its muscles and skin. The result is an oedemic, "puffy" chick that finds it hard to stand. It will eventually excrete the excess water but it is not a good start. Excessive weight loss from an egg will produce a dehydrated chick that will need access to water soon after hatching if it is to thrive – monitoring weight loss allows you to identify and care for these birds.

Monitoring weight loss is an effective way of ensuring that the hydration status of the chicks are optimised and so hatchability and quality are high. It does take some time and effort but it will reap rewards. I once advised a man hatching pheasant eggs. He was weighing the eggs on a regular basis and yet he couldn't get very good results. It turned out that he was achieving a weight loss of 19% at transfer (21 days) instead of the 13.5% he was supposed to get. I advised raising the humidity in his setters so that the correct rate of water loss was achieved and three weeks later the hatchability had risen by 10% of eggs set. In this case the person did not realise that he was at the wrong level but in other cases weight loss is recorded at the wrong time. Poultry hatchery staff often monitor weight loss at transfer (18 days) but at that point there is nothing they can do about any problems with incorrect weight loss. It would have be better to weigh eggs at 7 days and change the humidity accordingly so as to effect a change in weight loss. Collecting the data is only half the point about measuring weight loss – reacting to the results by adjusting humidity is essential to get the best results from your eggs.

Gas exchange

Aeration of the incubator is important because it introduces cool, fresh air that brings in oxygen and replaces the stale CO_2-rich humid air in the machine. The operator cannot alter the amount of fresh air entering many small incubators – the size and number of holes are fixed. Some of the larger incubators do have small flaps that can be slid over the holes. In general the amount of fresh air entering small machines will have little effect on temperature even with the holes fully open.

The oxygen supplied with the fresh air is crucial to embryo survival. Having said that the amount of oxygen needed by individual embryos is not great. For the first half of incubation embryos are small and their oxygen requirement is low but as growth takes off during the second half of incubation then oxygen demands increase as well. Recent comparisons of the rate of oxygen consumption of a wide variety of bird species has shown that there is no real difference in the metabolism of altricial and precocial species. Only the length of the pre-pipping plateau varies between species with smaller altricial species, typically with short incubation periods, having shorter plateau periods when oxygen consumption stops increasing. So long as fresh air is freely entering an incubator then there is no likelihood of problems with oxygen supply for any species.

Turning

Egg turning is critical for normal development but during artificial incubation it is usually taken for granted. Almost all machines that can be purchased all use an automatic system for turning. Most are based on the research on the effects of turning on poultry eggs but my recent research has shown that eggs from different types of bird require different rates of turning (see pp. 95–99). In view of this, the suitability of many table-top incubators for incubation of albumen-rich eggs could be questioned. Turning through 90° every hour may not be sufficient for many species. Indeed, additional hand turning has been shown to be beneficial for eggs from a variety of species including parrots, herons and cranes.

However, the more modern designs of incubators have incorporated systems to change the timing of turning events. As is shown elsewhere (pp. 97–98) eggs of species producing altricial offspring are turned more frequently than once an hour. As such, even though I do not know the turning rates of the majority of species, it can be assumed that eggs of species producing semi-altricial and altricial offspring will need to be turned more frequently. Research into poultry eggs has shown that a four-fold increase in the rate of turning had no detrimental effect and so increasing the turning rate of parrot, crane or penguin eggs (for example) to 3–5 times per hour will have no adverse effects on the development and will probably have a real benefit to the embryo. However, it may not be possible to increase the rate of automatic turning and additional hand turning may be necessary for some eggs. Rob Harvey has shown that this does have a benefit for many species. I hope that manufacturers will take heed of the research on turning rates of eggs from different species and allow operators of machines to adjust the turning rate. Ideally, they will also sponsor further work into this interesting field of embryonic development.

Eggs need to set appropriately to ensure that turning is correct. Eggs set in trays need to be placed with the blunt end upwards (Box 7.3, p. 125) to minimise malpositions (Box 9.3, p. 180) during hatching. If set in troughs or in between metal rods on moving base plates, then the eggs should be laid on their side and be free to move as the base plate moves beneath them. Sometimes after a few days of turning the eggs bunch up together and it is good idea to space them out again along the row. Restricting the eggs' positions in any way may cause problems with the orientation of the embryo within the egg and affect its ability to hatch (see pp. 94–95).

Turning is not needed all of the incubation period. Should the turning mechanism fail then eggs can be turned by hand until the incubator is fixed. Turning during the first third of incubation is most beneficial to normal development. Turning is not necessary in the hatcher. After transfer the eggs can lie on their sides and should be given room within the hatching compartment to allow for some movement as they roll around during the final stages of hatching.

Hatching

Hatching is the culmination of the incubation process and hopefully the incubation environment has allowed normal development of the embryo. As was shown previously (see pp. 63–65) the timing of hatching is not fixed for any species and you should seek to avoid the mistake once made by an novice Australian ostrich farmer. Having been told that the incubation period of the ostrich was 42 days he happily cracked open his eggs on the 42nd day of incubation and pulled out the embryos placing them in the hatcher to "dry off". After his barbecue he came back to find that all of his first hatch were dead. Whilst this is an extreme example it does illustrate the dangers of meddling with the natural processes of embryonic development. Temperature during incubation is a critical feature in the timing of hatching and only a couple of tenths of a degree can make several hours difference to the rate of development and will alter the incubation period. It is important that you are patient and you allow hatchlings to emerge of their own accord.

In small scale incubation it is often easy to candle eggs regularly (see pp. 157–159) to see whether there is movement in the air space of individual eggs. When the embryo can be seen moving in the air space the egg can be transferred into a hatcher. This machine is designed to hold eggs at the appropriate temperature and humidity but without turning (it is unnecessary during hatching). The lack of a turning mechanism usually means that cleaning of hatchers is relatively simple.

Some people will place a piece of corrugated paper or towelling on the floor of the cabinet to help the hatched bird stand up and to protect it from the metal surface. Some eggs that need to be "pedigree" hatched (*i.e.* the individual chicks have to be identified with each egg) are placed within smaller containers within the hatcher; empty margarine or ice-cream tubs lined with paper, or towel are useful in this respect. It is important that the egg is able to roll around relatively easily as this can help the chick to hatch.

In large capacity machines holding thousands of eggs the hatcher is a single stage machine and the temperature set point has to be reduced in order to optimise cooling of the eggs. In small table-top machines the cooling characteristics of the machines mean that there may be no need to lower the set point temperature from that used in the incubator. However, there is certainly no harm in lowering the set point temperature by 0.5°C, particularly if for large (100 g or more) eggs. It remains important in still-air machines that the temperature sensor is located at the top of the eggs. In force-draught hatchers the sensor has to be placed in the stream of air.

By contrast, the humidity in a hatcher generally needs to be high, *i.e.* 60–90%RH. This is to ensure that the eggshell membranes do not dry out after external pipping and hence trap the chick within the egg. Some people argue that high humidity is not necessary during hatching but there is no harm in it and its effects are likely to be beneficial overall. High humidity is usually achieved by having several water pans in the bottom of the machine, or by increasing the humidity setting on automatic systems. Of course the eggs themselves contribute to the higher humidity. Once the shell is broken water vapour can diffuse out of the egg without restriction. Weight loss by eggs in the period between pipping and full emergence of the bird can be 6–10% of initial egg mass. Having a high humidity in the cabinet reduces this rate of diffusion and prevents the egg and chick from dehydrating. Conversely, if the egg has failed to lose sufficient water by the time it goes into the hatcher then keeping it at a low humidity during hatching will not resolve the problem. The embryo will have processed the excess water and stored it in its muscles and skin and a dry hatcher environment will only add to its woes.

Small scale hatching doesn't really involve any modification of the gaseous environment within the cabinet, other than humidity. Provision of oxygen to the hatching eggs is critical and so air inlet holes should be fully open in table-top machines. In larger scale operations the amount of air exchanged is decreased once the eggs start to pip, which raises the humidity and the level of carbon dioxide in the air. Work with fowl eggs has shown that embryos and chicks are very tolerant of high levels of CO_2 in the air at the end of incubation and they can act as stimulus for external pipping.

Although it may prove difficult it is important to resist the temptation to help chicks to hatch. In particular, the chick has to break the eggshell itself and even though it may seem to be "having difficulties" it is best to let nature run its course. The level of CO_2 in the air space is critical in forcing the chick to break the shell. In the fowl egg, the CO_2 in the air space rises to 9% of the air and the bird is forced to externally pip to release this excess of gas (*N.B.* the high CO_2 may be uncomfortable for the embryo but it is not lethal in itself). Being impatient and making a hole in the air space doesn't make the chick hatch any earlier. Research on fowl eggs has shown that making a hole in the shell over the air space delays pipping but hatching occurs at the same time as control, unbroken eggs. Conversely, blocking the pores in the shell over the air space with wax brings pipping forward but has no effect on the time of hatching.

Those birds that are "having difficulties" are probably weak to start with and may not be able to hatch in the first place (see pp. 179–181). Whilst working on hatching of ostriches upon candling I often saw fully formed chicks that had broken into the air space but they failed to hatch. The reason was plain to see once the eggs were opened. The air space and the chicks were often covered by fungal growth and opening the air space would not have helped the bird and could have contaminated perfectly healthy chicks with fungal spores.

Some hatches will be over a discrete period of time, particularly if the eggs are set once or twice a week. This makes cleaning the hatcher out relatively easy because all of the eggs will be taken out the cabinet (whether they hatched or not). It is when eggs are set and hence hatch every day or two days, that keeping the hatcher clean can be problem. Hygiene in the hatcher cannot be over emphasised because failure to maintain high standards can lead to contamination and subsequent loss of chicks. When eggs are hatching all of the time, it is best to clean out the hatcher thoroughly every 3–4 days. If possible move eggs temporarily to another hatcher. However, if there is no alternative then leaving the eggs in a container loosely covered by a cloth in a warm room whilst the machine is cleaned should not pose too many problems so long as the time period is only 15–20 minutes. It is important that the hatcher is dry before eggs are returned to the cabinet (wet eggs chill fast) although it is okay if the eggs and cabinet can warm up together. Any containers that chicks hatch out in should be thoroughly cleaned, sanitised and dried before other eggs are put in them. Soiled paper should be discarded and other materials (*e.g.* towelling or plastic matting) should also be cleaned, sanitised and dried before being returned to the hatcher.

Artificial incubation – daily and weekly routines

It could be said that incubation is not the most exciting aspect of bird breeding. Except during hatching eggs don't do anything sitting as they do in a cabinet occasionally turning. Indeed I have always found it amazing that manufacturers are expected by their customers to put windows into incubators so that they can see the eggs. Incubation is much more a period of waiting for the excitement of hatching and a time when maintaining the incubation environment is very important. Therefore, good incubation practise invo-

Table 8.2. Checklists for routine monitoring and maintenance of incubators on a daily and weekly basis

Daily routine

Incubation parameter	*Procedure*
Power	Check that the machine is being supplied with power.
Temperature	Record the room temperature twice a day.
	Check and record the temperature reading on the incubator's display at least twice a day. Investigate any instances where temperature is not as set.
	In still-air machines ensure that the temperature sensor is located at the top of the eggs, particularly important if the cabinet has been opened to set or remove eggs (see pp. 120–122).
Humidity	Check whether there is water in the pans in the bottom of the machine or in the reservoir for automatic humidity delivery systems (see pp. 122–124). Replenish as necessary.
Aeration	Check that things placed on top of the machines do not block the aeration holes.
Turning	If possible ensure that the eggs are turning. Mark a cross on the top of a couple of eggs and record the position of the cross twice a day.
	If the turning mechanism fails then hand turn the eggs (see p. 124) until it is repaired.
	In machines where eggs are not restrained by egg trays, ensure that there is sufficient space between eggs and re-distribute any eggs that have "bunched-up" (see p. 152).
Hatcher	Candle eggs each day to investigate and record the progress of hatching (see pp. 157–159).
	Check whether eggs have pipped or hatched, recording the time events are seen.
	Remove hatched birds and clear up as much debris from the cabinet as is possible.

Table 8.2. continued

Weekly routine

Incubation parameter   *Procedure*

Egg candling (see pp. 157–159)	Check and record the progress of individual eggs noting the extent of growth of the vascular membranes or movements of the embryo.
	Note any eggs that may be infertile and remove once lack of development is confirmed over several days.
	Note any unusual black spots under the shell and remove any eggs which are clearly rotten (often smelly as well).
	Investigate the source of any smell of rotten eggs within the cabinet and remove any offending eggs.
Egg weighing	Record weights of eggs and calculate actual and predicted weight losses adjusting humidity supply according to results (see pp. 148–150).
Maintain hygiene standards	Ensure that hygiene standards established early in the breeding season are maintained. This can be done by having sufficient cleaning materials at all times and by discarding old disinfectant solutions in foot baths on a weekly basis.
Hatcher	Clean and sanitise the hatcher after each completed hatch or every week.

ves establishing routines for checking and operating the machines to ensure that embryos get the best chance of hatching.

Setting your machine up, putting the eggs in to incubate and leaving them until it is time for hatching is not an approach I would advocate. I have, therefore, tried to come up with some advice on the critical things that should be monitored and achieved on a daily, and on a weekly, basis and these are outlined in Table 8.1. It is important during incubation that problems are not only recognised but are then resolved as quickly as possible. Problems with temperature control or failure of the turning unit can have quite profound effects on results.

An important aspect of artificial incubation is good record keeping. This applies not only to the source and fate of eggs but also to the operating parameters of the incubator. Record the actual temperature and humidity being maintained in the machines, together with whether eggs are being turned. It is important to record the location of eggs particularly if you are moving them around between machines to optimise weight loss. Knowing when eggs are set and their anticipated date of hatching will help prevent problems with chicks hatching in the incubator rather than the hatcher. Unusual events (*e.g.* loss of power, unusual temperatures, humidity supply drying up, no turning) should be recorded in a diary because they may help explain any problems that may subsequently arise. A diary for each season is a useful reference and can be a good way of ensuring that procedures that brought success in one year can be repeated in the next or mistakes are avoided.

Apart from routine monitoring of incubation parameters, egg weighing and candling are the two most important procedures during artificial incubation. The former is described previously (see pp. 148–150) and the latter is described below.

Assessment of embryonic development

The eggshell not only serves to protect the embryo from physical and microbiological threats during incubation but it also prevents us seeing what is happening inside the egg. It is often useful to know whether an egg is fertile or whether it is developing normally but breaking the shell will ultimately kill the embryo. The problem is overcome by candling, *i.e.* shining a bright light through the shell to illuminate the interior of the egg.

I'm not sure when candling was invented but the name derives from the practise of using wax candles to illuminate the egg in a darkened room. Nowadays, electrical halogen bulbs often provide the light but a bright torch can also be used. Candling is best carried out in a dark room as this improves what can be seen within the egg but it does depend on the eggs. White and other pale-shelled eggs are a dream to candle revealing a lot of detail. More heavily pigmented shells tend to block the light more and spotting on the shell tends to confuse the view. Turkey eggs are also relatively heavily pigmented and commercial hatcheries use specialist ultraviolet candling lamps to see inside the eggshells. Many eggs, particularly of cranes, bustards and emus have heavily pigmented shells that block penetration of the light into the egg irrespective of the brightness of the light. UV lamps may help with candling such shells. The advent of emu farming led to the development of infrared candling lamps to see inside the dark green shells. The infrared lamp is placed on the shell and the light transmitted by the shell is picked up by a separate sensor and displayed on a screen.

Candling shows us whether the embryo is developing – the blood vessels of the yolk sac and chorio-allantoic membranes can be seen as shadows under

158

Figure 8.8. Relationship between the percentage of eggshell covered by dark shadows observed during candling and the percentage of the incubation period.

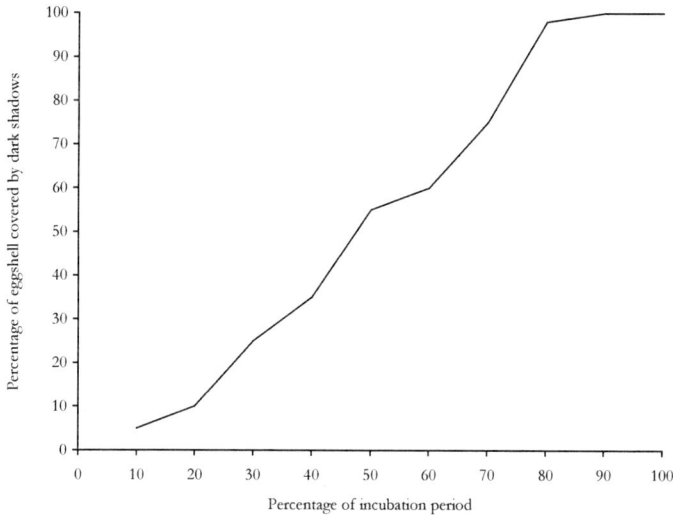

the shell. In some instances the embryo can also be seen and during the second half of incubation movement is often observed. The lack of membranes or shadows usually mean that the egg is infertile. The pattern of expansion of the extra-embryonic membranes follows a typical course (Figure 8.8). For the first half of incubation there is rapid expansion of the shadows as the yolk sac membrane enlarges. The rate of increase in the size of the membrane area then slows dramatically at around 50% of the shell area because the yolk sac membrane grows under the yolk and away from the shell. In the meantime the chorio-allantoic membrane has been growing over the top of the yolk sac membrane and the rate of increase in the shadows increases again once the chorio-allantois grows beyond the limits of the yolk sac membrane (Figure 8.8). Most of the shell is covered by 80% of the incubation period. If the pattern of shadows does not change over time then this usually indicates that the embryo has died.

Candling can also reveal problems with microbial contamination. Bacterial colonies form of the inside of the shell membranes and are seen as black spots, although they are more difficult to see in spotted eggs. Eggs that are black throughout during the first two-thirds of incubation are usually addled and will smell bad. Whenever an egg is suspected of being going bad it should be removed from the machine to prevent it oozing any material or exploding within the cabinet.

Recognising infertile eggs, patterns of development of microbial contamination by candling takes a little practice, particularly with pigmented or thick shells. The simplest way to learn is to write down what you see during candling for individual eggs. Notes on how much the shell is covered by membrane or the general appearance of the contents can be useful to consult when an egg is candled again a few days later. Eggs that have shown no change tend to be infertile or dead.

Summary

- Egg quality is important for successful hatching.

- The best way of combating microbial contamination is through good hygiene at all stages of breeding and incubation.

- Hatchery design is important to reduce cross-contamination and separating incubator and hatcher rooms is ideal.

- Eggs can be stored easily but the viability of the embryo can be diminished if environmental conditions are poor.

- The incubation environment needs the appropriate air temperature and humidity, mixing of fresh and incubator air and egg turning.

- Monitoring weight loss during incubation is highly advantageous.

- Hatching best takes place in a hatcher.

- Daily and weekly routines for monitoring incubation are suggested.

- The role and importance of candling are described.

9 – What to do When Things go Wrong

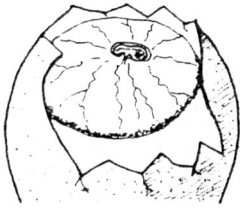

It is unfortunate but not all eggs that we incubate go on to successfully hatch. Failure of normal development can be attributed to a wide variety of problems that are related to the egg itself, how the egg was handled or how it was incubated. Trying to decipher the clues left by undeveloped or dead embryos is far from simple and it is easy to make mistakes in the diagnosis of problems. In commercial incubation there are thousands of chicks hatched and thousands of dead eggs. Opening all of these eggs would be an impossible task and so only a sample can be studied in order to establish a pattern of mortality. When there are only a few eggs involved then each dead egg can be investigated in the hope of trying to determine why individual embryos died and to see whether there are any broader patterns of mortality that may indicate a more general problem.

How to approach this task? There are two main options to choose from. I could have discussed problems based on the time at which they are observed during development but I felt that this potentially becomes over complicated. For instance, embryos could die in the last 24 hours before hatching because of a vast variety of reasons, which include: malpositioning, temperature was too high or too low, humidity was too high or too low, eggshell conductance was inappropriate for the humidity used, shell-induced hypoxia, lack of turning during the first half of incubation, microbial contamination leading to yolk sac infection, the egg was cracked during transfer to the hatcher,..... I could go with the list but I feel that it is not a constructive approach. Instead I have chosen to describe the problems associated with specific aspects of breeding or incubation. In this way I can focus on those factors that are most likely to cause a problem and thereby provide a checklist of things to investigate first before going onto the more exotic, and perhaps more difficult to resolve, problems.

Before proceeding it is important to realise that not all embryos are capable of developing normally. Early embryonic mortality is a key feature of avian and mammalian reproduction but there is a big difference. If a mammalian embryo dies early in development then it can be absorbed by the body or expelled and the mother probably doesn't even register the pregnancy. By contrast, birds' eggs are laid irrespective of whether they are fertile, or any embryo is viable or not. If the embryo dies within a couple of days of the start of in-the egg remains as a monument to its failure. In this way I feel that eggs lend hope into the hearts of people incubating them – there is the anticipation that they will hatch even if it is false hope. Embryos can die for reasons that have nothing to do with the things that people have done to them and it has to be accepted as a part of nature. The number of embryos dying early is not large. In commercial poultry and game birds embryonic mortality at 5 days of incubation is typically 3–4% of eggs set and I guess that this would apply equally to other species. It is only when levels of early mortality are much higher that they are indicative of another problem that requires further investigation.

Whilst trying to temper the enthusiast's yearning to hatch every egg it is also true that there should be an expectation of hatching a high proportion of fertile eggs. Too many people accept low levels of hatchability because their expectations are set too low. At least 80–90% of fertile eggs should hatch under artificial incubation. I believe that there is no intrinsic reason why artificial incubation should be any different to natural incubation; it is perhaps only our inability to provide the correct incubation conditions for individual species that lowers hatchability. The following sections attempt to highlight the commonest problems with artificial incubation. I hope to provide a comprehensive analysis of problems but it is inevitable that I be unable to cover all eventualities. However, it is important to check out the obvious things first and only when incubation conditions are shown to be perfect should other potential problems be considered. Let's start with information on how to recognise dead embryos.

How to recognise dead embryos

Not all eggs contain embryos and so it is important to be able to recognise infertile eggs. This is not that difficult, particularly if the eggs have not been incubated for a long time. Infertile eggs have the same appearance as unincubated eggs. The easiest way to see this for yourself is to buy some fowl eggs for eating and crack them open into a bowl. The yolk is typically uniformly coloured and there will be a small (1–2 mm diameter) white spot on the upper surface. If you can't see this dot then wait awhile and it will float up to the top of the yolk. This dot is the "blastodisc" and indicates that fertilisation did not take place. Under a microscope the blastodisc looks like Swiss cheese – full of holes rather than a flat uniform plate. The yolk around the blastodisc is typically uniform yellow and not discoloured. An exception is shown by eggs opened after a long (i.e. many days) incubation period when the yolk can take on a mottled appearance with dark patches over its surface.

If the yolk is discoloured, particularly if there is a white "skin" over the yolk surface then it is likely that the embryo started to develop but died early in incubation. There is often an accumulation of watery, paler sub-embryonic fluid (it resembles skimmed milk) under this layer. A recognisable embryo is usually not seen in the white layer and if this layer is about the size of a penny or a cent piece (20 mm diameter) then the embryo probably died around day 1–2 of incubation. Older embryos are often seen in association with blood vessels (which may be brown in colour) and the extra-embryonic membranes and volume of sub-embryonic fluid can be much greater. Often the membranes are the most obvious structures and an embryo is not recognisable. This doesn't mean that the embryo wasn't there but in the time between it dying and the egg being opened the embryo disintegrated. In such instances the embryo has died around day 3–4 of incubation. Older embryos are larger and so tend not to degrade that quickly. They are more prominent with black pigmented eyes clearly seen. In general, small embryos measuring less than 20% of the egg in length have died during the first third of their developmental period.

Embryos dying between 3–6 days are often accompanied by a "blood ring" around the yolk. This is caused by the nucleated red blood cells sinking down and accumulating in the terminal vein running around the rim of the yolk sac membrane. Normally the beating of the heart keeps the cells evenly distributed. This is often red but can be brown particularly if the egg is opened late in the incubation period.

Embryos that die during the second third of incubation have defined bodily structures and wings, legs and toes, are clearly visible. These embryos are larger than younger embryos but still remain small (less than half the egg's length). The older embryos are typically fully feathered and appear to be small birds. Growth of the extra-embryonic membrane is extensive and the chorioallantoic membrane lines much of the inside of the shell.

Embryos that die in the last third of development are large and progressively fill the eggshell. Size is usually the best criterion for ageing these embryos – the bigger they are the older they are. Closer to hatching the egg contents become progressively drier as first the allantoic fluid and then the amniotic fluid are absorbed. Precocial species have fully feather embryos during the last third of incubation but altricial embryos remain naked except for tufts of down. Relatively large embryos that have not pipped into the air space are often surrounded by these fluids prompting people to say that they had "drowned". This is not possible – the embryos never had air in their lungs in the first place. Rather the embryos have died some time before internal pipping perhaps of hypoxia (see pp. 169–171). It is often useful to make distinction between those embryos that died after internal pipping (the beak is in the air space) or after external pipping (the beak has broken the eggshell). Embryos preparing for hatching have adopted the typical hatching position (see

Table 9.1. Broad characteristics of dead eggs. For more details see text (pp. 161–163)

Infertile	Appearance of unincubated egg, uniformly coloured yolk and blastodisc measuring 2–3 mm in diameter.
Died first third of development	Embryo non-existent or measuring less than 20% of the egg's length. Accompanied by extensive growth of extra-embryonic membranes and "blood rings" are commonly seen.
Died second third of development	Embryonic limbs and beak are clearly seen but embryo is still small (< half the egg length). Extra-embryonic membrane development is extensive with a large amount of the eggshell covered by the chorio-allantoic membrane.
Died last third of development	Embryos are bigger, bird-like and filling the egg towards the end of development. Egg contents may be dry due to loss of fluids. Late-term embryos will have retracted their yolks and adopted the hatch position if not internally or externally pipped.

pp. 57–58) and have started yolk sac retraction (see pp. 59–60) and have a lot of their yolk within their abdomen.

For simplicity assigning an age of development for dead embryos is best achieved by saying whether they died in one of the three thirds of incubation (summarised in Table 9.1, p. 163). In this way a pattern of mortality can be established that can be used to help diagnose problems. In general embryonic mortality during the first third of incubation is relatively high, it is low during the second third and very much higher during the last third. You would normally expect to see a peak in late mortality that was around twice that for mortality during the first third of incubation.

Fertility

The most obvious cause of failure of development is infertility of the eggs. Infertile eggs are nothing to do with egg handling or incubation conditions. Rather the proportion of infertile eggs laid by a female is a function of the mating frequency (or rate and efficiency of artificial insemination [Box 9.1, p. 164]). It is always important to ensure that eggs removed during candling as "infertile" are indeed infertile. It is not good enough to assume that fertility assessed by candling is a measure of true fertility. High rates of early embryonic mortality indicate other problems with egg handling or incubation conditions but infertile eggs are a symptom of problems with breeding birds.

For instance, in pheasants a mating ratio of 12 females to one male pro-

BOX 9.1 – ARTIFICIAL INSEMINATION (AI)

Not all birds are able or willing to mate naturally and so techniques for artificial insemination have been developed in a range of species ranging from ratites to birds of prey and bustards. These involve milking the male bird for semen. Usually the bird is imprinted on people and sexually stimulated by a person. The semen is then collected during this false mating. Semen contains millions of spermatozoa and so can be diluted before being manually inserted into the vagina of the oviduct.

The advantages of AI lie in being able to spread the sperm from a particularly good male around a number of females. It is often used in conservation projects where male birds may be in short supply. In the turkey industry it is used to produce fertile eggs because natural mating between commercial turkeys has become difficult. Semen can also be stored frozen for future use.

Disadvantages include failure of insemination due to poor techniques. Semen can also be contaminated by faeces or urine rendering it ineffective. Some females may not be responsive to the handling involved during AI. Fertility may be a little erratic until appropriate techniques are developed for novel species.

duced a fertility value some 4.5% lower than birds kept at a ratio of 8 females to each male. In other species of birds forming more long term pair bonds mating may not be taking place in the first instance but the female is still stimulated to produce eggs. Putting two birds in a cage doesn't mean that they "like" each other and are sexually compatible. I found this with farmed ostriches. If females could choose their mates then fertility was high but if birds were simply put together then mating was not guaranteed. Indeed, if there were two females to each male, then fertility was very low and only increased substantially once one of the females had been removed.

Note that it is usually the female that initiates and allows mating in birds. The male has to impress the female to gain her confidence and mating is not guaranteed even if the pair has been together for a long time. More importantly, some birds respond to people as sexual partners and may ignore their own species once the people have gone. Again in ostriches I showed that both males and females displayed to humans and birds often mated in front of people standing at the fence line. Unfortunately, this only showed that the male was able to take advantage of the female's attraction to the person. When observed out of the ostriches' sight these same birds would assiduously avoid each other. Contact with humans is often useful in taming birds for AI (Box 9.1, p. 164) but imprinting can occur without any deliberate attempt by the people. In many species being captive bred for release into the wild, strict measures have to be taken to prevent imprinting. Conservation programmes for condors and cranes often use model heads to feed hatchlings and visual contact with people is avoided at all costs.

In small scale operations, where good record keeping should identify the parents of each egg, it is simple to identify birds that consistently produce infertile eggs. It may be necessary to change pairs around to see whether there are more productive partnerships. If a bird continues to fail to perform then it may have to leave the breeding programme.

One problem with assigning fertility to eggs is the fact that embryos can die whilst they travel down the oviduct and are apparently infertile because they do not show any development during subsequent incubation. In these circumstances, more modern techniques to determine fertility could be used to assess whether mating is efficient. The number of spermatozoa trapped in the peri-vitelline membrane or the number of holes made by sperm in the peri-vitelline membrane (see pp. 46–47) can be used to assess whether an egg was fertile. This is a rather destructive technique and would only be applicable to eggs that were deemed infertile after a few days of incubation.

Incubation temperature

Maintaining the correct temperature is the most crucial aspect of artificial incubation and so it is first thing to be checked if you have a problem during incubation. Get temperature right and things will probably go well but get it wrong and your chances of hatching healthy chicks are greatly reduced.

Most problems with incubation temperature are due to setting the up machine incorrectly. Still-air incubators need to have the tip of the temperature probe at a level equivalent to the top of the eggs (and all eggs have to be about the same size). In force-draught incubators the probe has to be located in the air stream and not in a "dead spot" in the cabinet.

Alternatively, the temperature probe may be at fault. Glass thermometers can develop splits in the column of mercury or alcohol that extend up the graduated scale. A small bubble forms in the liquid and so the thermometer can read a different temperature than its is actually registering. As a result the machine can be set at an incorrect temperature. This problem is usually resolved by cooling the thermometer down in a refrigerator – this usually draws all of the fluid down into the bulb and the column will reform once the thermometer is warmed up. To be accurate electronic temperature probes need to be properly calibrated to the correct temperature. Problems arise when the calibration "drifts" and the probe fails to register the correct temperature. This problem is usually resolved by reference to a high quality glass thermometer, which allows the electronic probe to be re-calibrated.

Figure 9.1. Effect of weekly exposure to different incubation temperatures during the first (grey columns), second (white columns) and third (black columns and dashed line) week of incubation on percentage of embryonic mortality for domestic fowl eggs. From Romanoff & Romanoff (1972).

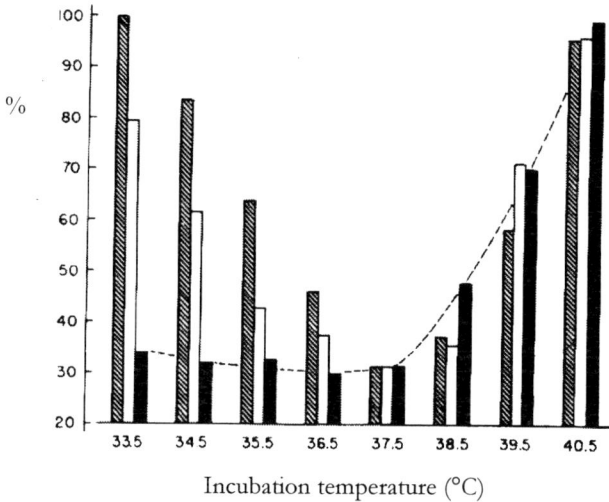

Incubating at too low a temperature causes slow rates of development and extends the incubation period. Note that small changes in temperature have profound effects on the rate of development. A machine set at 37.0°C will take longer to hatch eggs than a machine set at 37.5°C; for the domestic fowl the time difference will be approximately 12 hours. Embryos that are developing slowly are usually still alive at the time that they should have hatched and if left alone and given more time, they should hatch in good numbers. The bigger the departure from the normal incubation set point then fewer embryos will develop normally. Deformed toes and feet are commonly seen in eggs incubated at low temperatures (<35°C). It matters when the embryos are exposed to low temperatures. As is shown in Figure 9.1, there is a lot of embryonic mortality following 7 days of incubation at temperatures of 33.5 or 34.5°C during the first or second thirds of the incubation period. By contrast, the impact of these incubation temperatures is much lower for embryos exposed to these temperatures during the last third of incubation.

Periods of cooling lasting a few minutes are common in a wide range of species that leave their eggs unattended during natural incubation (see pp. 71–78). There is little evidence to suggest that embryonic mortality is adversely affected by brief periods of cooling in a nest or in an incubator. In the fowl younger embryos seem to be more sensitive to periods of cool temperatures than older embryos but I know of no other studies that investigate whether this is true in species with altricial or semi-altricial young. Hence, the few minutes spent candling eggs will not cause any great problems with embryonic viability. I know of commercial mallard producers who, starting on day 14, cool their eggs for 30 minutes a day (and spray them with water before replacing them in the machine) but the reasons why this should be important have not been investigated. Other people are known to periodically cool their non-poultry eggs by turning off the heater, removing the top of the machine or opening the doors. Unfortunately whether regular, brief periods of cooling are in any way beneficial to embryonic development in an incubator remains unclear and requires sound scientific investigation before it can be confidently recommended for any particular species.

Exposure to higher than average temperatures are much more of a problem to embryos. Although a small increase in incubation temperature (1°C) will cause a decrease in the incubation period, any further increase will not produce shorter incubation periods but rather the incubation period will increase (see pp. 142–144). A decrease in temperature of only 2°C will not cause too much mortality but a rise of the same magnitude will cause high embryonic mortality irrespective of when the embryos are exposed (Figure 9.1). A one degree increase in incubation temperature (*i.e.* 38.5°C) for turkey embryos causes massive increases in malposition 5 (head in the small end [see Box 9.3, p. 180]), subcutaneous haemorrhages (bleeding under the skin), excess albumen proteins and swollen (oedematous) heads. I have observed similar symptoms in fowl and pheasant embryos in machines that were operating at higher than optimal temperatures. I would not be surprised if these symptoms are also seen in non-poultry species.

Excess albumen is commonly seen in overheated embryos. Unlike the unused albumen seen in unturned eggs (see p. 172), this albumen is a honey coloured sticky fluid that covers the body of the embryo. It appears that in warmer than normal eggs the embryo runs out of time to swallow all of the albumen in the amniotic fluid (see p. 56) and as it prepares to hatch it still has albumen proteins covering its body. This is the reason why those chicks that hatch from overheated eggs have sticky down.

Incubation humidity

If temperature is within acceptable limits then humidity is the next most important aspect of incubation where problems can arise. The usual problem is caused by not getting the humidity correct for the porosity of the eggshell. As was pointed out in Chapter 8 (see pp. 145–148) it is critical that the appropriate humidity is employed to ensure the correct rate of weight loss. Some of the problems associated with incorrect humidity are based around problems with eggshell quality (see pp. 174–177).

Eggs with large air spaces and dried up contents are associated with excessive rates of weight loss. This may be due to one of two reasons. Firstly, the porosity of the eggshell is high and the humidity is insufficient to prevent too much water loss. Thin shells and cracked shells are often in this category. The problem may be resolved by raising humidity early in incubation. The second possible reason is that the humidity was too low for the eggshell porosity, which is in the normal range. Raising the humidity is required to rectify this problem.

Embryos in eggs that have not lost enough water during incubation are often puffy (oedematous), with swollen legs and abdomen, and there may be a lot of free fluid within the shell. Again the humidity could be at fault with a normal shell being kept in a humidity that is too high to produce a normal weight loss. Adjusting the humidity conditions will help remove this problem. Alternatively, the humidity may be normal but the eggshell has a low porosity. In this situation the lack of weight loss is more a symptom rather than the problem in itself. Embryos that fail to hatch from low porosity eggshells tend

---

BOX 9.2 – INCUBATION AT ALTITUDE

Birds breed in all manner of places including high latitudes (towards the poles) and at high altitudes up mountains. This latter situation poses a real problem to the embryos because at high altitudes the atmospheric pressure is lower, the partial pressure of oxygen is reduced and gases diffuse faster. This reduces the partial pressure difference across the eggshell and there is a danger that the embryos will not get enough oxygen for normal growth and survival.

The birds respond to the high altitude by having relatively high values for conductance of their eggshells, which allows more oxygen to diffuse into the egg. Unfortunately, this also allows more water vapour and carbon dioxide to diffuse out. Artificial incubation has to ensure that the level of oxygen in the cabinet air does not fall by having a high rate of air exchange but this also causes a loss of humidity. As high rates of water loss are deleterious to the embryo it is important to monitor weight loss from the eggs and maintain an appropriate (high) level of humidity. It is the restriction in the supply of oxygen that will cause the greatest problem to the embryo.

If the breeding population of birds are kept at altitude then the chances are that they will have adapted their eggshell porosity to the prevailing conditions. This will not be so for any eggs laid at lower altitudes and transported up mountains for incubation.

Figure 9.2. Effect of eggshell porosity (measured as water vapour conductance) on the oxygen consumption of domestic fowl embryos measured on day 15 of the 21 day incubation period. The diagonal line indicates the maximum oxygen consumption allowed at each value of shell porosity.

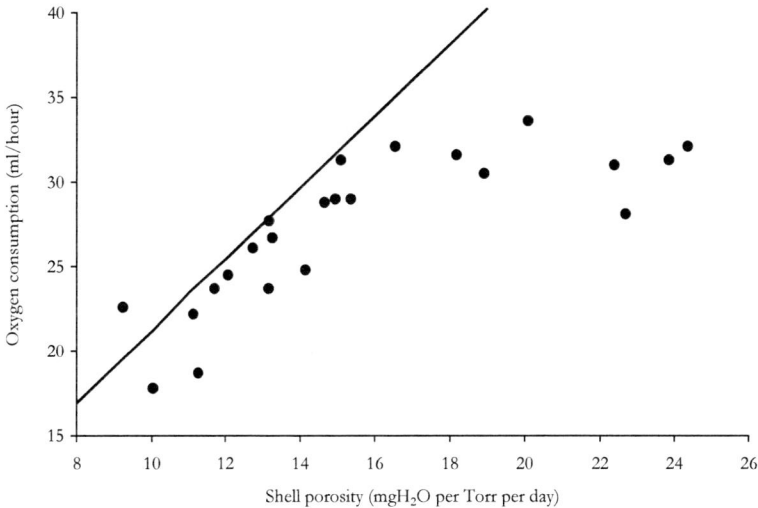

to be dying through lack of oxygen rather than the problem of losing insufficient water (see pp. 169–171). Lowering the humidity will allow the correct rate of weight loss to be achieved but this will have no effect on the amount of oxygen entering the eggshell.

Respiratory gas exchange

Bird embryos need oxygen to survive and this is provided by diffusion of this gas through the eggshell pores (see p. 35). Embryos face problems when there is reduction in the amount of oxygen available to them. This could be due to either or both of the following situations.

Firstly, the amount of oxygen in the air decreases. This is typically due to the embryos actually using up the oxygen without any fresh air being available to replenish the gas supply. Normally the oxygen in fresh air is at 21% of the total gas (a partial pressure of 148 Torr). Failure to allow air changes will cause the oxygen concentration of the air to decrease. This causes a reduction in the difference in oxygen concentrations either side of the shell and so less oxygen can diffuse into the egg. A natural situation where this can occur is at altitude (See Box 9.2, p. 168). Lack of oxygen reduces rates of growth and will kill embryos in the days approaching hatching.

The other situation is where the eggshell has a low conductance to oxygen and so won't allow enough oxygen across. Unfortunately, the embryo cannot lower the amount of oxygen in its blood indefinitely and there is a lower limit of around 100 Torr. Prolonged periods below this level are not possible and the embryo will die of lack of sufficient oxygen. There is a natural variation in eggshell porosity and so some eggshells on the lower part of the range do not allow enough oxygen to cross the shell. This is shown in the oxygen consumption of domestic fowl embryos (Figure 9.2). Above the average porosity value the shell does not restrict the amount of oxygen that can diffuse across and so the embryo can take as much oxygen as it wants (Figure 9.2). By contrast, oxygen consumption for lower than average porosity eggshells is restricted by the eggshell and is below the maximum amount of oxygen that can diffuse into the egg (Figure 9.2). This same situation has been recognised in domestic turkeys and ducks and is seen relatively early in development. The lack of oxygen retards the growth rate of embryos in lower porosity eggshells during the last third of incubation. In extreme cases they end up so small that the embryos are unable to hatch normally and succumb to the hypoxic conditions (*i.e.* they suffocate due to the low levels of oxygen within the shell) before they internally pip into the air space. Embryos in low porosity eggshells in a poorly aerated incubator are hit twice and so will die early in the growth phase.

These types of dead embryos are recognised by being small despite appearing fully formed. They are surrounded by a full complement of membranes and often there is a lot of allantoic and amniotic fluid in the egg. They rarely internally pip. Often described as "drowned" these embryos have never breathed air and so cannot drown although they do suffocate in the egg.

Resolving the problem of porosity induced hypoxia is relatively simple. All you have to do is to ensure that sufficient oxygen is supplied to the embryo at all times and so the incubator and hatcher should have good supplies of fresh air at all times when there are eggs without externally pipped eggs. The rooms in which machines sit should also have good aeration and ventilation to ensure that the room air does not get stale. Under these circumstances most embryos will suffer but survive the problem of a low porosity eggshell. Only embryos in extremely low porosity eggshells will suffer so much that they die.

It is important to note that getting the weight loss correct does not mean that the embryo will get enough oxygen. Although water vapour and oxygen move through the same pores their diffusion rates are independent of each other. Incubation under low humidity can mean that a low porosity eggshell can attain the correct rate of weight loss but it will not affect the amount of oxygen in the air or the amount entering the egg.

The only exception to having a high rate of aeration is in a hatcher where a batch of eggs of the same developmental age have started to externally pip. In this circumstance you can close the aeration holes to raise humidity and CO_2

levels. Those chicks that have externally pip can use their lungs to breathe the rarefied air and suffer no adverse effect. The high partial pressures of CO_2 stimulate those eggs that haven't externally pipped to do so and so the spread of hatch is reduced.

Shell induced hypoxia is usually recognised by monitoring weight loss – the low porosity eggshells will have a low weight loss under average conditions. Only weighing early in incubation will identify these eggs so that a small hole can be made in the eggshell over the air space, which may help alleviate this problem a little. Such holes have to be made well before hatching so as to allow sufficient oxygen to enter the egg to allow for normal growth as the metabolic rate of the embryo increases. Moreover, the rate of weight loss has to be carefully monitored and the highest hygiene standards are required to prevent any airborne contamination of the air space by fungal or bacterial spores (see pp. 173–174).

Aeration also has a function of diluting carbon dioxide produced by the eggs. CO_2 is a waste product of the embryo's metabolism and is potentially toxic to the embryo. However, domestic fowl embryos are very tolerant of high levels of CO_2 (Figure 9.3). Even though embryos are most sensitive to CO_2 during the early stages of their development they can still tolerate 1% CO_2 without any embryonic mortality and require 7.5% CO_2 to cause complete mortality. At the end of incubation the embryos can tolerate 7% CO_2 without ill effect and levels of 17% CO_2 are need to kill all of the embryos (Figure 9.3). This tends to put the normal levels of CO_2 recorded in incubators of 0.3–0.5% into context. As with a lot of things about incubation this work has only been done on domestic fowl embryos but there is little to suggest that other birds will be affected differently. Indeed recently the patterns of metabolism in a wide range of precocial and altricial species have been shown to be remarkably similar.

Levels of CO_2 have been noted in the past, and present, because of the ease by which this gas can be measured. By contrast, it is difficult and expensive to measure levels of oxygen in the air. High levels of CO_2 suggest low levels of O_2 and so they are a useful measure of problems with machine aeration. Most table-top incubators and hatchers will not experience many problems with accumulation of CO_2 so long as there is an adequate amount of fresh air going through the cabinet.

Egg turning

Turning rarely gives a problem in precocial species unless there is a malfunction of the equipment. However, the symptoms of lack of turning are pretty clear – hatching will be late and hatchability will be depressed. The co-

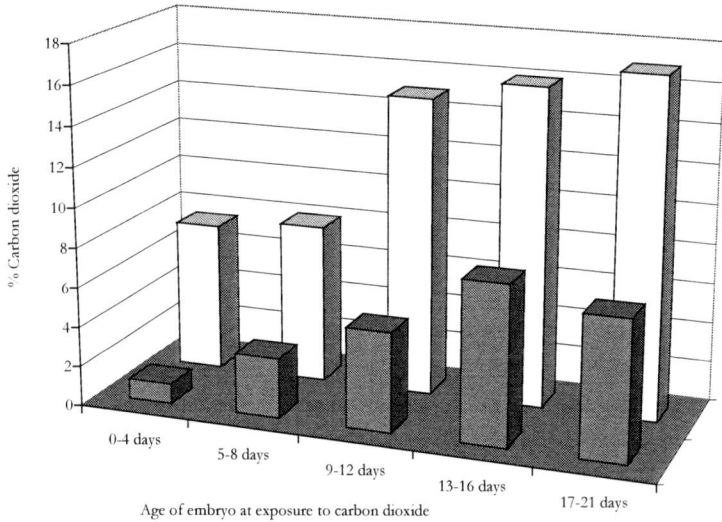

Figure 9.3. Tolerance of domestic fowl embryos at different ages to levels of carbon dioxide in the incubator air. Grey columns indicate percentage of CO_2 where there is zero embryonic mortality and the white columns indicate the CO_2 percentage that causes 100% embryonic mortality.

mmon idea that unturned embryos simply stick to the shell membranes and die early in incubation is largely unfounded in most birds. Although there will be more early dead the biggest mortality occurs in the last few days of incubation. These dead embryos will be small, have failed to internally pip into the air space and there will be a lot of residual albumen. Rather than lying over the embryo's body (*c.f.* elevated incubation temperatures, see p. 166), there will be quite a lot of albumen lying in the bottom of the egg and it is covered by the chorio-allantoic membrane. Malpositioning of the embryo (Box 9.3, p. 180) will also be increased. These symptoms are very characteristic of the lack of turning during incubation.

Little work has been done on the effects of insufficient turning during incubation of non-poultry eggs and in particular albumen-rich altricial eggs. Studies have shown that these eggs are turned more frequently in natural incubation (see pp. 95–98). It is likely that the symptoms of unturned eggs described above will also occur in these eggs. If all other possible problems have been discounted for a particular group of eggs not hatching then lack of turning should be seriously considered. The frequency, angle and the regularity of turning of altricial species is still a matter of much debate and require further research.

Microbial contamination

An obvious cause for the loss of eggs during incubation is microbial contamination. These eggs are recognised by their bad smell, they may ooze material from the pores, they might explode upon being picked up (so-called "bangers"), or upon opening the contents are unrecognisable. Often the contents are black and quite firm. Either way they will smell awful. Other eggs have the appearance of cream cheese. If the contents are emptied out the shell then bacterial colonies can often be seen growing on the inside of the shell membranes. Bacterial and fungal (hair-like structures) colonies are often observed on the inner shell membrane lining the air space.

The organisms infecting eggs are nothing unusual and reflect the normal microbial flora found in the nesting environment and in bird droppings. Hence, common bacterial types are *E.coli*, *Pseudomonas*, *Proteus*, and *Salmonella*. Common fungi are *Aspergillus* and *Candida*. These organisms are easily seen growing in the eggs because they form relatively large colonies but mycoplasmas and viruses do not form discrete colonies and are harder to detect.

The anti-microbial properties of albumen (see pp. 38–40) seem to prevent a lot of minor infections under the shell from getting any worse but eggs can go bad very quickly after the embryo has started to develop. Once the microorganisms come into contact with embryonic tissues they can easily kill and degrade the embryo. Fungal growths tend be slower to develop and many embryos seem able to survive through to hatching but they usually eventually succumb to the problem.

Microbial contamination is a key feature of poor nest hygiene (see Box 8.1, p. 130) and most problems arise within a few seconds of an egg being laid into a dirty nest or on to the floor (see Figure 8.1). Good hygiene should prevent contamination of other eggs during handling and incubation although poor egg washing procedures can be real source of microbial contamination. However, allowing addled eggs to ooze fluid (the build up of waste gases in the eggshell creates tremendous pressures that can force fluid through the microscopic pores in the shell) in the incubator can cause real problems of cross-contamination with uninfected eggs. Bad eggs should be identified (their smell usually lets you know they are there) and removed from the machine before they can ooze any fluid. If there is fluid loss, particularly if the egg explodes in the cabinet (or anywhere else for that matter), then the machine (and the surface of any contaminated eggs) has to be cleaned thoroughly and disinfected as soon as possible.

I have recently identified an important problem with microbial contamination. Yolk sac infection is a common problem in chicks soon after they have hatched and I suspected that this could also be a problem in the egg. To study this I swabbed the yolk sacs of embryos that had died in the egg without externally pipping the shell. The subsequent cultures showed that late term embryos in broiler fowl eggs collected from nests had a contamination rate of 10% of the sample but eggs collected from the floor had contamination rates of 30%. In pheasants eggs laid in wire-cages had a yolk sac infection rate of 24% of late-term embryos. However, it is more typical for pheasant eggs to be laid on the ground in a field and these eggs had contamination rates of 60% of the sample of embryos. This equates to 7–8% of fertile eggs set dying of yolk sac infection within 36 hours of hatching! Development of a nest box that keeps the eggs clean and would be used by the birds would be a real innovation in game farming. These results just reinforce the idea that the best way of tackling microbial contamination is by prevention of the problem.

Resolving problems of microbial contamination may be difficult if they are associated with eggs laid outside of clean nests. However, maintaining good hygiene during breeding and incubation will assist in preventing the problem becoming too great. If results are poor due to microbial contamination it is worth getting swabs cultured to identify the organism(s) involved and to isolate the source of the contamination.

Eggshell quality

It has to be said that not all eggs that are laid are worthy of being set to incubate. There can be problems with extensive contamination with faeces or soil (see pp. 130–132) or the eggshell itself is defective. Assessment of problems in eggshell quality can be rather subjective with some examples of shells being acceptable to many people and unacceptable to others. Here I would like to offer some suggestions for reasons why eggs should be downgraded on the basis of their shells.

Eggshells with a bulge, often most easily identified by touch during handling, should be deemed as rejects because the swelling in the shell represents damage caused during egg formation. Some bulges can be very apparent (Figure 9.4A) whereas others can be quite small. Normal shell deposition is around the yolk and dehydrated albumen proteins wrapped in two fibrous shell membranes. As the albumen proteins are "plumped", *i.e.* expanded by absorbing fluids, the volume of the albumen increases and stretches the shell membranes; the calcitic shell is then deposited on the taut shell membranes. The egg also revolves within the oviduct and so a symmetrical shell shape develops. A bulge in the shell represents a defect in this process probably associated with damage to the membranes, or perhaps the shell, early on in calcite deposition. The resulting egg is not symmetrical and is often infertile, a result perhaps of direct contamination of albumen by the fluid used to deposit the calcitic eggshell.

Figure 9.4. Examples of poor shell quality in domestic fowl eggs. A) shell with bulge on left side; B) slab-sided shell; C) rough, "chalky" thin shell; D) wrinkled shell at sharp pole; E) wrinkled shell all over;

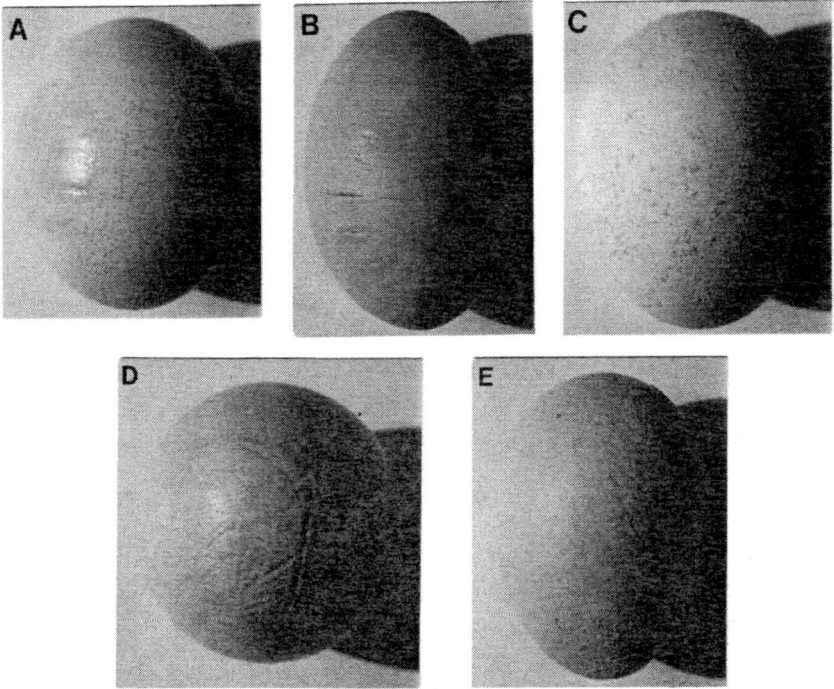

Slab-sided shells (Figure 9.4B) have one side that has been depressed during shell formation and the deformation is made permanent. Such problems are probably by physical assault to the hen. It seems that any damage to the integrity of the shell during formation appears to prevent normal development thereafter and these eggs are often found to be infertile.

Eggs not completed before oviposition have chalky shells, which are pale in colour, thin and rough in appearance and touch (Figure 9.4C). These eggs often crack when picked up but if set for incubation, the process of handling them usually means that they will not hatch. The thin shell allows excessive moisture loss during incubation, and these eggs are easily damaged during handling and cracks increase moisture loss further.

Wrinkled shells represent problems during deposition of calcium carbonate in shell gland. Some shells have wrinkles at the sharp pole (Figure 9.4D) and these can be characteristic of an individual hen. Such eggs are usually worth setting but eggs with extensively wrinkled shells (Figure 9.4E) are not. It is likely that the hens had been under a degree of stress during egg formation and there is a problem with watery albumen and so the shell is rough and wrinkled in appearance. Fissures in these shells can be points of weakness in the thin shell and so such eggs are easily damaged.

Repair shells (Figure 9.4F) represent an extreme situation where the shell has been broken as it was being formed and it is repaired during the normal process of shell deposition. Such problems are caused by stress, or more likely, a physical insult to the bird during shell deposition. Any embryos in these eggs rarely develop perhaps due to contamination of the albumen by fluid rich in calcium.

Eggs with white or purplish chalky deposits on the outer surface have an additional layer of calcium phosphate deposited on top of the cuticle when these eggs are retained in the shell gland for too long. The additional deposits tend to block the pores in the shell causing the eggs to lose less moisture dur-

Figure 9.4 continued. F) repaired cracked shell; G) Shell covered with white deposit of calcium phosphate; H) impact crack; I) hairline crack; and J) shell with hole in side.

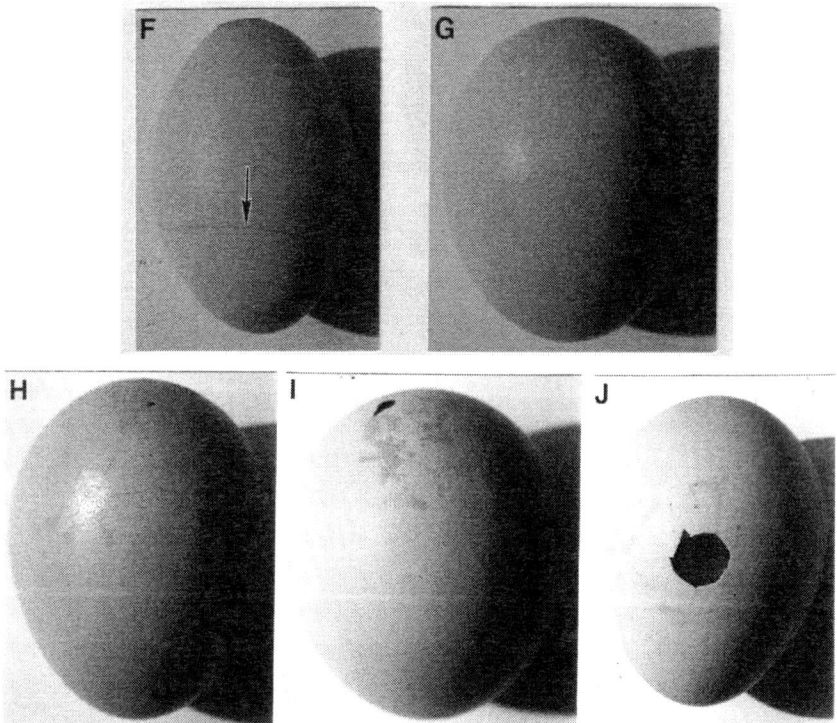

ing incubation than they would have. Lower porosity eggshells prevent oxygen movement into the egg, which considerably reduces the chances of the chick hatching (see pp. 169–170). Extensively marked shells (Figure 9.4G) should be rejected as hatching eggs.

Although it can be sometimes difficult to identify them, eggs with cracked shells (Figure 9.4H & 9.4I) should not be set because they will not hatch. They are prone to rapid dehydration of the yolk and albumen in the first few days of incubation causing the death of the embryo. Cracked eggs are also prone to high rates of microbial contamination which is often established prior to the eggs being collected. As well as causing loss of the egg, contamination can cause problems with 1st grade eggs which are contaminated by direct contact with rotten eggs, or with contact with rotten egg contents when these eggs explode (see p. 173). There is no point in setting eggs with extensive cracks or holes in the shell (Figure 9.4J). They dehydrate within hours in the setter and can rot easily.

Very small eggs almost certainly represent eggs that have developed from yolks that were released from the ovary before they were mature. This means that they are often not suitable for fertilisation. Some small "eggs" don't even have a yolk and consist solely of albumen surrounded by a shell. Obviously they are infertile and are not worth setting. Very large eggs often represent those containing two yolks. Although these may go on to develop two fraternal twin embryos (*i.e.* they come from two separate blastoderms) the likelihood of them hatching is very remote.

The shape of some eggs is critical with regard to hatching. Abnormally shaped eggs may not fit in the egg trays or allow normal development to proceed. In particular, if torpedo-shaped eggs are abnormal for the species then they may not provide sufficient room for the chick to rotate in as it tries to hatch. Very round eggs do not sit well in square holes and can sometimes fall off eggs trays when tilted.

Setting eggs upside down is also a problem when they are fixed in one position (see Box 7.3, p. 125). Recognising which end of the egg has the air space may not be that in easy in some species but after a few days of incubation the air space will be larger and easier to see. Once identified the egg should be set in the correct position to maximise its chance of hatching.

Handing and storage of eggs

Rough handling of eggs has been shown to be a sure way of killing embryos. Shaking eggs before incubation increases the incidence of malformations. Shaking during incubation causes damage to the extra-embryonic circulation or development of nervous tissues leading to very high mortality particularly during the middle third of incubation. Jarring eggs (by dropping a heavy weight on to the table containing the incubator) during the first 12 hours of incubation caused complete mortality. Hitting eggs without cracking the shell causes moving air spaces to form, which prevents normal development through to hatching.

It is true to say, however, that there is a physical side to natural incubation. Being sat upon with movement due to turning and changes in the position of the incubating bird mean that eggs have evolved to be tough and the embryo is resilient to such physical treatment. Birds are not necessarily gentle in their manipulation of eggs and this only becomes a problem when eggs are not what they should be. In the past problems with pesticide accumulation in birds of prey caused thinning of eggshells and breakage of the shell under normal incubation conditions was the typical problem (rather than, for instance, changes in eggshell porosity). It is probably the regular nature of experimental jarring or shaking of eggs that cause problems although floating air spaces can develop spontaneously.

My advice is to treat eggs like bone china – delicate and fragile and yet strong enough to resist the physical aspect of their time in the nest. Even though birds may be quite rough with their eggs in principle it is much better for us to treat eggs with utmost respect all of the time.

Washing is a major procedure often carried out on eggs before they are set. Incorrect washing can cause real problems with microbial contamination and in some directly affects embryonic mortality. The biggest problems occur if the washing solution is at the wrong temperature or has been allowed to get too dirty so that it no longer has any sanitising effect. If the solution is too cold then the egg may be warmer than the solution and air in the pores may contract pulling in the washing solution. This may directly kill the embryo or it may introduce micro-organisms into the contents. This problem is exacerbated if the solution has been used to wash too many eggs and the disinfectant action has been used up. In this instances mortality is going to be high early in incubation.

Having the washing solution too warm or leaving the eggs in the solution for several minutes can lead to the egg contents absorbing and retaining heat from the solution. This not only reduces the temperature of the solution or any subsequent eggs but it may raise egg temperature above the physiological zero for embryonic development (see pp. 140–143). This may cause embryos to develop beyond the gastrula stage and their ability to withstand storage is compromised.

This "pre-incubation" can also occur if eggs are left for long periods in warm conditions before being collected. Experiments have shown that the ideal cooling rate of a fowl egg is about 4 hours to go from body temperature to 18°C. Obviously bigger eggs will cool more slowly and smaller eggs more quickly. If fowl eggs remain at relatively high temperatures (above 25°C) for any length of time then embryonic development can be initiated. Putting such eggs into storage thereafter reduces their ability to survive once true incubation starts. This problem also arises when storage conditions are not ideal, *i.e.* the temperature is too high or it fluctuates so that embryos are exposed to warm temperatures for significant periods of the day.

In nature some eggs survive extended periods of time under conditions that we would consider completely inappropriate. For instance, the first ostrich eggs in a clutch are left for 3 weeks before incubation proceeds and yet they survive the fluctuating temperatures of a warm day and cold night in tropical Africa. We do not fully understand how this is achieved and there is certainly a need for more research into storage of eggs from non-poultry species. In the meantime, under artificial conditions it is best to maintain storage conditions that are stable, cool and relatively humid (see pp. 136–138).

In many species embryos have reached the gastrula stage of development at the time of egg laying. In poultry this has been shown to be the ideal stage of development for the embryo to survive storage. Work with poultry has shown that holding eggs at low temperatures for prolonged periods causes the blastoderm to shrink and to take longer to initiate further development once the correct incubation temperature has been achieved. Therefore, eggs stored for 10 days will develop slower than fresher eggs.

Storage is a problem because it reduces the viability of the embryos and so mortality is often increased during the first few days of incubation. Mortality during the middle of incubation is also increased. Those embryos that survive are often retarded in development and will hatch late. The further the storage conditions depart from the optimum (see p. 136) then the bigger the problem will be.

Hatching

Many embryos die during hatching but many of the problems they experience are due to situations that arise earlier in incubation. They may die in the hatcher but conditions in the hatcher did not kill them. There are a couple of specific problems that arise during the hatching period. However, the process of hatching is physically demanding and its success largely depends on whether the embryo has adopted the correct hatching position (see pp. 59–62) and has grown to sufficient size.

The most obvious problem is the embryo adopting an incorrect position for hatching and are "malpositioned" (see Box 9.3, p. 180) upon *post mortem* examination of the egg. Unfortunately, the factors that cause malpositioning are very poorly understood. Often "bad luck" on the part of the embryo is probably the cause – the embryo just finds itself in the wrong place at the

BOX 9.3 – MALPOSITIONS DURING HATCHING

The embryo has to adopt a specific position for it to hatch normally. The neck is bent around with the head under the right wing and the beak pressed against the inner shell membrane. If the embryo adopts a different position it is called a hatching malposition. There are six common types (see below [illustration by the great embryologist Alexis Romanoff]).

Malposition 1: Head *above* the right wing.
Malposition 2: Head under *left* wing.
Malposition 3: Position normal but beak away from the air space.
Malposition 4: Leg over head.
Malposition 5: Head at the small end of the egg.
Malposition 6: Head between legs.

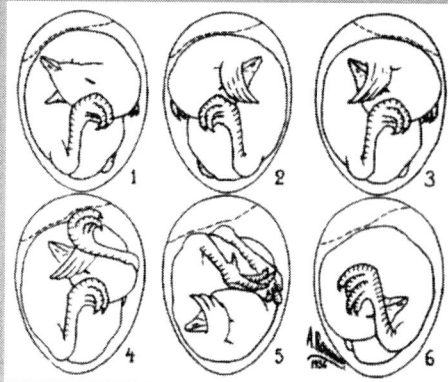

The reasons why these malpositions occur are not full understood. The easiest to comprehend is malposition 5, which is the usual consequence of setting eggs upside down but it can be a spontaneous event but this may relate to overheating during incubation (see p. 165). A deficiency of vitamin B_{12} in the egg is associated with a high incidence of head between the legs (malposition 6). Lack of egg turning also causes a high incidence of malpositions.

Having the head away from the air space can prevent internal pipping and make hatching harder to achieve. Having the leg over the head can restrict movement within shell but it is unclear why having the head over the wing rather than under wing makes such a difference to the embryo. Similarly the reason why having the head under the left wing does not allow normal hatching is not clear.

wrong time. However, problems with turning and female nutrition are implicated in increasing the incidence of malpositions in poultry species (see Box 9.3, p. 180). Malpositioning is not due to the egg being held in the wrong position within the hatcher. It is true to say that eggs should have a degree of freedom to roll around within the hatcher as this often assists in the embryo's manoeuvres during hatching.

One problem I saw with ostrich chicks hatching out was associated with the rotation of the bird within the shell as it hatches. The chicks were rotating without a problem but failing to break the shell as they went and so all you saw through the original pip hole was the back of the bird. If not recognised early enough the chick died. This was probably due to suffocation as the beak was well away from the hole and so the bird probably could not get enough oxygen to survive. If the beak was not visible in the pip hole then cracking the shell around the egg's equator did help some of the chicks to hatch without a problem.

One real problem during hatching is the dehydration of the shell membranes after external pipping. Once the shell is broken the egg can rapidly lose water vapour through the hole and if the hatcher environment is too dry then the shell membranes can dry out, stick to the down or skin and trap the embryo within the shell. Although some people may argue that it is not necessary there is certainly no harm in maintaining a high humidity in the hatcher. It can help reduce the risk of membranes drying out.

Hatching is a critical stage of a bird's life and so it should be left to its own devices. Too many people are keen to assist in the process and this can cause too problems. Making a pip hole for the chick doesn't really help it that much and can release CO_2, a valuable stimulus to pipping. More importantly, if the egg is contaminated with fungi or bacteria then making a hole may release spores into the air that may contaminate other clean eggs or chicks. Whenever possible leave eggs to hatch on their own – the chick will be of better "quality" than if it is pulled out of the shell.

Female nutrition

When all other aspects of incubation have been shown to be within acceptable limits then the composition of the egg contents is one of the other factors that may be considered as a potential problem affecting hatchability. This problem is a function of the diet of the female laying the eggs – to a large extent eggs are what the bird eats.

This point is illustrated well by the fatty acid composition of the lipids in egg yolk. Fatty acids are small molecules that are the building blocks of the larger fat molecules deposited by the bird in the yolk. The exact breakdown of the various fatty acids depends a lot on the fatty acid composition of the diet.

Table 9.2. Symptoms exhibited by poultry embryos under a deficiency of vitamins or minerals in the egg.

Vitamin	Deficiency symptoms
A	Early mortality from failure to develop circulatory system. Abnormalities of the kidneys, eyes and skeleton.
D	Late mortality with soft bones and defective upper mandible.
E	Early mortality with haemorrhaging and failure of circulation.
K	Late mortality, no clear diagnosis.
Thiamin	High mortality during hatching, no easy diagnosis but high rates of polyneuritis in surviving birds.
Riboflavin	Peaks in mortality early, mid-term and late stage, altered limb and mandible development, dwarfism and clubbed down.
Niacin	Bone and beak malformations if niacin formation (from tryptophan) is prevented.
Biotin	Late stage mortality, parrot beak, chondrodystrophy, skeletal deformities and webbing between toes.
Pantothenic acid	Mid-term mortality, subcutaneous haemorrhaging.
Folic acid	Late mortality, bent tibiotarsus, syndactyly and malformations of mandible.
B_{12}	Late mortality, atrophy of legs, oedema, fatty organs, haemorrhaging and head-between-legs malposition.
Manganese	Late deaths, dwarfism, short long bones, head malformations, abnormal feathers, oedema.
Zinc	Late deaths, defects of the spinal column, missing limbs, poor development of the eyes.
Copper	Deaths at early blood stage.
Iodine	Prolonged hatching and incomplete abdominal closure.
Iron	Poor extra-embryonic circulation, low blood haematocrit and haemoglobin.
Selenium	High mortality early in incubation.

BOX 9.4 – MALFORMATIONS

There are times when embryonic development takes a wrong turn but the embryo does not die immediately. The result is a malformed embryo. The degree of the anatomical defect can vary considerably but usually involve the brain, eyes or other parts of the nervous system, the head and beak, the heart or skeleton. Separate or conjoined (Siamese) twins are also considered as a malformation.

Many malformed embryos are spontaneous and without any specific cause. Other problems may be inherited or are associated with specific nutritional deficiencies (*e.g.* poor skeletal growth is associated with a deficiency of vitamin D). Some malformations are associated with the incorrect temperature or humidity during incubation.

It is quite easy to get concerned about malformed embryos, or even chicks (birds with exposed brains are relatively common in commercial poultry), but they should only be a cause for concern if they occur at high levels. In general, malformed embryos are rare – in poultry I have seen deformed embryos at incidences of 0.2–0.5% of eggs set and I would only get worried about high numbers of malformed embryos (perhaps above 2% of eggs set).

Hence, you can now buy fowl eggs for eating that are rich in poly-unsaturated fatty acids, supposedly a healthy aspect of the diet. These eggs are only possible because the hens were fed a diet rich in these molecules.

The fat composition of yolk has recently become a matter of interest because birds of different diets, and hatchling maturities, are being studied. Hence, yolk composition varies between species (see pp. 40–41) and within some species, *e.g.* geese, pheasants and ostriches, it has been shown that the diet of the females affects the fatty acid profiles. Those females on a natural diet produce yolks with a different fatty acid profile than birds fed solely on a commercial diet. Unfortunately the effects of these differences on embryonic development and hatchability have yet to be fully investigated. The possibility certainly exists that the different composition does affect embryonic viability.

Other key components of bird egg yolk are vitamins and trace minerals. The embryo relies on the female to provision the egg yolk with all that it requires for development. Experimentation and commercial mistakes have demonstrated that deficiencies in vitamins and minerals can severely curtail normal development. The typical effects of a deficiency of various vitamins and minerals are shown in Table 9.2.

It has to be said that studying whether female nutrition is a critical aspect of determining failure of artificial incubation is very difficult. In general, embr-

Table 9.3. Brief summary of the symptoms of problems during hatching together with possible causes of the problem. See text for further explanation.

Symptom	Possible cause of problem
No sign of embryonic development	Lack of fertilisation
Long incubation periods - embryos are alive at end of normal incubation period	Low incubation temperature, or eggs are being chilled
Long incubation periods - embryos are dead at end of normal incubation period	High incubation temperature
Embryos covered in honey-coloured sticky fluid or chicks with sticky down	High incubation temperature
Large air space, dried out contents	Low humidity leading to excessive weight loss, poor shell quality, or cracked shell
"Puffy" embryos or chicks	Humidity too high causing low weight loss
Lots of fluid around well developed but small embryo	Hypoxia due to low shell porosity or low rates of aeration
Small fully formed embryos with residual albumen lying in the bottom of the egg covered by the chorio-allantoic membrane	Insufficient egg turning
Microbial growth, bad smell, substances on shell surface	Contamination by micro-organ-isms usually soon after egg laying
Fully formed embryos with discoloured yolks	*In ovo* yolk sac infection
Floating air space	Probably rough handling
Slow development and high early embryonic mortality	Prolonged pre-incubation storage

yonic mortality will follow a similar pattern irrespective of the parents and the incubation conditions need to shown to be optimal before diet should be considered. Studies would need to establish that two different diets actually have differing effects on hatchability and chick survival.

The best approach is to prevent problems arising as far is possible. This is achieved by avoiding reliance on a single source of nutrition. Many commercial diets are advertised as complete rations but the danger lies in the chance that the formulation of the diet is not suitable for the species involved. This point is illustrated for the fatty acid composition of the rations - game bird rations are often based on poultry formulations and do not necessarily provide the correct balance of lipids for the birds. It is much better to provide a variety of food items, particularly fresh vegetation and meat, and to supplement the diet of birds with vitamins and minerals. As with all captive animals (humans included), birds will thrive on a mixed balanced diet.

Genetics

It is not always the case that the embryo is viable. The problem may be genetic in origin and relate to the genetic make-up of the parents of the potential offspring. In poultry, some genes in individual embryos have been identified that are lethal to embryos. Furthermore, mutations of genes are often the cause of congenital malformations (Box 9.4, p. 183) and these are increased by inbreeding of stock. Such problems require considerable detective work and breeding experiments to identify and the simplest way to counteract genetic problems in reproduction is to do as much out-breeding as is possible. Avoid breeding closely related birds too many times and introduce unrelated stock into the breeding population as much as possible. On the whole whilst genetic problems remain a possible explanation for embryonic mortality, it should be only be considered seriously once all of the other options discussed above have been discounted. The exception to this is where a very small pool of genetic stock exists, as in the case of conservation of endangered species. Here genetic problems due to inbreeding are a major issue that is difficult to resolve without careful management of the existing breeding stock and the offspring.

In conclusion

Failure of precious eggs to hatch can be extremely frustrating. This feeling of frustration is often compounded by a lack of an explanation why the eggs failed to hatch. The wide range of possible causes of an individual embryo's death makes trying to resolve incubation problems often quite difficult. My approach is to look for wider patterns within the problem that may reveal what is going wrong. The simplest approach is to try to assess when the embryos died – if indeed you have any embryos at all. Once you know this then you can start to investigate the possible causes of death. It is important not to focus on individual embryos and worry about big problems when the solution may be quite simple.

For instance, you have a still-air incubator and it will not hatch eggs of one particular species. The embryos are dying at the end of a long incubation period. This would suggest that the temperature is set too high. It may not be a fault with the machine, particularly if smaller eggs from a second species hatch perfectly well in the same machine. Perhaps you have set the temperature to the correct level for small eggs but this does not suit the other type and the embryos are being kept at too high a temperature. In this case using different incubators for the two species will probably rectify the problem.

Always start with the obvious ones – fertility, incorrect temperature, incorrect humidity, lack of oxygen, no turning or microbial contamination. Some of the most obvious symptoms of common problems are summarised in Table 9.3. If all of the obvious aspects of incubation are acceptable then you need to look at other aspects of breeder bird management or incubation procedures that could be affecting embryonic viability. Trying to resolve these problems may be much more difficult.

Summary

- Factors that may adversely affect the viability of embryos are considered but it has to be realised that some embryos will die irrespective of the incubation conditions.

- How to recognise infertile eggs and the approximate developmental age of dead embryos is described.

- Infertile eggs can be a common problem when eggs fail to hatch.

- Incubation temperature is the first thing to check if there is a widespread problem of high embryonic mortality.

- Incorrect weight loss and lack of oxygen are also important.

- Incorrect turning may be an issue with eggs from altricial species.

- Microbial contamination can kill embryos at any time during incubation.

- Eggshell quality can cause problems with maintaining the integrity of the egg during incubation or with rates of weight loss.

- Pre-incubation handling can damage eggs and reduce the viability of embryos.

- During hatching embryonic mortality is often associated with malpositioned embryos or with dehydration of the shell membranes.

- Embryos that die whilst in the hatcher are usually not killed by the conditions in the hatcher but have been exposed to conditions in the incubator that reduced their chances of hatching.

- Female nutrition and genetics can adversely affect hatchability but problems in these areas may take time to resolve.

10 – Artificial Contact Incubation

Previous chapters have described the conditions created by birds and people for the successful incubation of eggs. My hope is that you will have seen the similarities and differences between the two types of incubation system. However, as a summary I wish to clarify these points in order to highlight why people have had to operate and design incubators in the way that they currently do. In this chapter, I then go on to speculate on whether natural or artificial incubation is best and what factors need to be considered. A new concept of artificial incubation is then described in light of these ideas.

Natural and artificial incubation: similarities and differences

It is relatively easy to breakdown the aspects of the incubation systems employed by birds and people into their component parts and these are summarised in Table 10.1. For this I have compared incubation in nests with artificial incubation in small table-top incubators. This is primarily because of the smaller numbers of eggs that are incubated in small incubators, which are comparable to natural clutches of eggs. There are other factors that need to be considered when dealing with large scale machines but these are not considered here.

Temperature control in a nest is through direct contact between the eggs and the bird. Heat is transferred both ways by conduction and the volume of air in a nest under the bird is generally small. After all, the bird doesn't want to waste energy heating lots of air. This system creates a temperature gradient within the egg with the top part of the egg being warmest because of its proximity to the brood patch. The magnitude of the gradient depends on egg size (see Figure 5.10). As incubation proceeds metabolic heat production and development of embryonic circulation act together to greatly reduce temperature gradients in the egg (see pp. 83–85). In a machine heat transfer is by convection and relies on good movement of air over the eggs. The volume of air in the cabinet is generally large because this helps to act as a buffer in temperature regulation as well as providing standing room for hatchlings.

Humidity around eggs in a nest is created by water vapour from the eggs and skin of the bird being trapped in the small amount of air surrounding the

Table 10.1. Comparison of the incubation parameters of contact incubation by birds and artificial incubation in table-top incubators.

Factor	Natural incubation	Artificial incubation
Supply of heat	By conduction via contact with the brood patch of the incubation adult.	By convection from air heated artificially.
Removal of metabolic heat	By conduction via contact with the brood patch of the incubation adult. Access to cooler air in the nest especially when adult bird stands up.	Cool air is heated by warm eggs.
Egg temperature	Temperature gradient in eggs with hottest part at the top adjacent to the brood patch.	Temperature gradient in eggs in still-air incubators. No temperature gradient in force-draught machines.
Volume of air around eggs	Minimal.	Relatively large.
Humidity	The bird-nest incubation unit raises humidity above ambient but there is no direct control.	Humidity is typically controlled by adding water to supplement humidity from eggs and ambient conditions.
Aeration	Exchange of air through nest walls as well as periodic changes of air when bird rises to turn the eggs or to leave the nest.	Continuous.
Ventilation of eggs	Diffusion when bird rises from the nest.	By convection in still-air machines or by a fan in force-draught incubators.
Turning	Periodic but random in frequency and angle.	Periodic but regular in both frequency and angle.

eggs. Humidity is raised above ambient but usually fluctuates as ambient conditions change (see Figure 5.15). Artificial incubation invariably aims to add water vapour to raise the humidity to a level found to achieve a suitable weight loss during incubation.

Although there can be exchange of air through the nest walls, the extent of this depends on the materials used in the construction of the nest. In general, fresh air is introduced into the nest by the bird rising from the eggs either to turn the eggs or to leave the nest to forage. Diffusion and convection rapidly replace the warm air. In an incubator, air is supplied continuously through holes in the cabinet and provides a key source of cooling as well as bringing in oxygen and diluting waste gases, such as CO_2 and water vapour, produced by the eggs.

Turning by birds is a haphazard process where the frequency and angle that the eggs move are often randomly distributed. Individual eggs may not be moved in any particular turning event. On average the eggs of precocial species experience a turning event once an hour but altricial species turn their eggs much more frequently. In artificial incubation turning is regular and has historically been based on research on precocial poultry species.

These differences in incubation control have led some people to question whether conventional convection-type incubators are the best system for incubation of all types of bird egg. Although natural incubation systems are likely to be the best system to employ for many instances, I do not believe that there should be any disadvantages to artificial incubation. The need to use bantam fowl to incubate eggs of other species for the first half of incubation does not suggest that the birds are ultimately better. Rather it shows that there is some fault in the system of artificial incubation that is prejudicing normal development in a machine. Those problems that arise can often be attributable to problems with the design or operation of the incubators. Moreover, some birds are poor parents under captive conditions.

Poor hatching results have led some people to abandon incubators and employ systems that mimic the natural situation. This often involves warming the eggs from above using warmth from hot water bottles or electrically heated blankets, thereby attempting to recreate the temperature gradient that an egg would experience in a nest. Whilst people have had notable success with this system there is relatively little control over factors like gas exchange and humidity, and turning has to be done by hand. It would be better to mimic the environmental conditions experienced under a bird but with the environmental control of an artificial incubator.

An artificial contact incubator

For many years Frank Pearce has thought about the differences between natural and artificial incubation. He has long held the belief that the temperature gradient that develops across an egg under a bird is really quite important in the development of the embryo. Evidence to support this idea came from important studies by Scott Turner and results of still-air incubators, which seem able to produce good hatchability for a variety of species despite rather vague environmental control. Hearing of recent successful experiments in the USA, where incubation was attempted using hot-water bottles, led Frank and the team at *Brinsea Products Ltd* in England to develop a new system of artificial incubation that uses direct contact to conduct heat into the eggs.

Briefly the artificial contact incubator (Figure 10.1), which is protected by world-wide and US patents, is a cross between still-air and force-draught machines. The machine is a table-top incubator holding ~80 domestic fowl eggs. The top of the cabinet is closed on the underside by a thin plastic or rubber film that can be inflated with air that is thermostatically controlled to an appropriate temperature (38.0°C). A fan within this "bubble" ensures that there is uniform distribution of warm air. The plastic film is pushed down to mould around the top half of the eggs in the same way as the eggs make contact with a brood patch. There is no other source of heat in the rest of the machine. Hence, the top of the egg is warmed by contact with the heated air across the plastic sheet and a temperature gradient develops across the egg. Later in incubation, when metabolic heat production is high, eggs can lose some of this energy by helping to warm the air in the top of the machine.

Figure 10.1. An illustration of the Brinsea Products contact incubator. See text for more details.

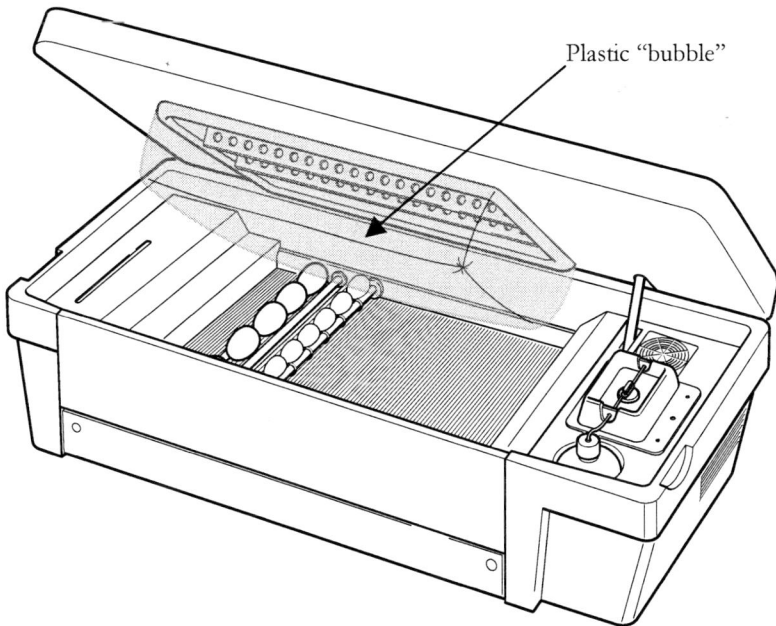

For turning, eggs rest on plastic rollers (Figure 10.1) that are rotated automatically by a conveyor beneath. The intervals between consecutive turns, the angle of each turn and indeed the direction of consecutive turns can all be pre-set to specific values or be chosen at random by the micro-controller. Just before turning happens the air in the top of the machine is pulled out raising the plastic film off the top of the eggs allowing them to move freely. This process also sucks in cool air through filter medium and relatively large holes in the floor of the cabinet. The filter medium can be chosen to mimic the porosity of a natural nest and provides the only means of aerating the part of the cabinet holding eggs. After turning is completed the air is blown back into the top of the machine and the plastic film makes contact with the eggs again. Humidity is supplied from water-soaked pads lying a platform below the moving base plate. The system is under micro-processor control.

Preliminary results using twelve prototype machines for incubation of eggs from a variety of non-poultry species have been highly encouraging. The system appears to work well and hatchability in the new machines matches, if not betters, that of eggs incubated in conventional incubators. If it continues to prove to be a successful system what will be the reasons for the success?

Two particular aspects of the machine operation are different from conventional machines. Firstly, there is a temperature gradient across the eggshell, which is absent in a conventional force-draught incubator. Of course this mimics natural incubation but a gradient is also found in a still-air machine. The use of a plastic film gives more control of temperature regulation at the top of the egg thereby removing a source of error in a conventional still-air system where problems are often caused by the incorrect positioning of the probe controlling temperature.

However, the new machine also differs from a still air machine in an important way. The raising and lowering of the plastic film during turning events aerates the eggs by exchanging most of the air with cool fresh air. In the still-air system, air exchange is continuous through loss of warm air via convection through the top of the cabinet (see pp. 115–116). In this system the rate of air exchange is slow. This massive exchange of air in the new machine mimics the effect of the bird getting up off the eggs and allowing fresh air to replace the warmed air in the nest. This achieves a relatively rapid cooling of the eggs.

It is unclear which of these two factors will prove to be most critical in the operation of this type of incubator. Further research and trial work will be needed to clarify this point. However, a practical artificial contact incubator does seem to be a viable option for incubation of bird eggs and this may overcome many of the problems associated with convection-type machines.

In conclusion

On this interesting note for development of artificial incubation in the 21st century I have come to the end of my description of incubation by birds and in machines. Clearly, we understand a lot about natural incubation and how bird behaviour brings about the environment suitable for embryonic development. Over the past century, our understanding of artificial incubation has also matured and hatchability of poultry eggs in commercial incubators can be very high. However, the reliance on information about poultry species for our understanding of incubation of wild birds may not all be good. The recent research reported on the rates of eggs turning in nests in a wide variety of species highlights this point. We may have been missing a critical aspect of normal incubation that promotes good development by assuming that all eggs require to be turned at the same rate as poultry eggs. Users of small incubators will need to take into account the idea that the eggs from more "exotic" species of bird may need to be turned more frequently. Brinsea's new contact machine allows turn intervals to be set as frequently as every five minutes.

Research into both natural and artificial incubation continues but not to the same extent as in the past. This does not show that we know all that is to be known but perhaps reflects more on financial constraints for funding research in general. I wish it were not the case but the challenges faced by artificial incubation in the next 25 years will come from a need to develop captive breeding programmes for increasing numbers of bird species. There is certainly an urgent need for the scientific study of artificial incubation of non-poultry species, especially for birds that lay eggs that produce semi-altricial or altricial offspring. Increased understanding of natural incubation and the application of this knowledge to design and operation of artificial incubators should help to improve results for all types of birds.

Summary

- Natural and artificial incubation systems are compared.

- A novel system that bridges the gap between contact incubation by birds and convection incubation in a machine is described.

- Research into natural and artificial incubation remains relevant in the 21st century although there should be an increased focus on species laying eggs producing semi-altricial and altricial offspring.

Glossary

Abdominal cavity – Part of body containing the internal organs.

Acid-base balance – Physiological balance between acidity and alkalinity in the blood and tissues.

Acrosome – Top part of head of spermatozoon containing enzymes needed in penetration of the membranes around the ovum.

Aeration – Introduction of fresh air.

Air sac – Membranous, air-filled extension of lungs.

Air space – Cavity formed between the outer and inner shell membranes as air enters the egg to replace water vapour lost through the pores during incubation.

Albumen – Egg white, comprising mainly of proteins; comes in two forms: thin and thick.

Allantois – Membranous sac growing from gut of embryo which acts as the embryonic bladder.

Altitude – Height above sea level.

Altricial – Born naked and helpless.

Ambient – Surrounding.

Amnion – Membrane growing out from embryo's skin to form protective sac.

Anti-microbial – Prevents growth of, and kills, microbes.

Artificial incubation – Incubation of eggs using machines.

Avidin – Protein found in egg albumen that binds the vitamin biotin.

Bacteria – Single celled, microscopic organisms, many capable of producing disease.

Bactericidal – Kills bacteria.

Biosecurity – Security of life usually against disease causing organisms.

Bladder – Membranous sac used to store urine.

Blastoderm – Embryonic stage at oviposition.

Blastodisc – Circular cellular structure in infertile egg at oviposition.

Brood patch – Bare patch of skin on abdomen which is rich in blood vessels and is used to transfer heat from the bird to eggs under incubation.

Broodiness – Desire to sit on eggs.

Brooding – Protect and keep warm.

Calcite – Crystalline form of calcium carbonate ($CaCO_3$).

CAM – See chorio-allantoic membrane.

Candling – Procedure of shining a bright light through an eggshell in order to see inside.

Capacity – Number of eggs that can be set within an incubator or hatcher.

Carbon dioxide – A gas forming 0.03% of normal air produced as waste product during cell metabolism.

Chalazae – Twisted albumen proteins that help to suspend the egg yolk in the albumen.

Check thermometer – Monitoring device to check air temperature in incubators.

Chick – Newly hatched bird.

Chick fluff – Pieces of downy feathers, which break off during hatching.

Chorio-allantoic membrane – Combination of chorion and allantois to form one membrane that lies adjacent to the inner shell membrane.

Chorio-allantois – See chorio-allantoic membrane.

Chorion – Outer extra-embryonic membrane surrounding embryo.

Chromosome – Thread-like structure composed of DNA and found in pairs in the nucleus of every cell. Contains genetic information.

Cloaca – Cavity into which gut, ureters and reproductive ducts open.

Clutch – A batch of eggs produced by a particular bird for one bout of incubation.

CO₂ – Carbon dioxide.

Cold spot – Part of incubator which is cooler than rest of machine.

Condensation – Formation of liquid from vapour.

Conductance – Measure of the porosity of an eggshell to a gas.

Conduction – Transfer of heat through physical contact.

Cone layer – Inner most layer of calcitic eggshell which is attached to the outer shell membrane.

Contact thermometer – Thermometer which incorporates metal electrical contacts, which through the presence or absence of mercury in the thermometer's column are linked to heaters and cooling pipes, thereby allowing temperature control.

Contamination – Impurity, especially by microbes.

Convection – Heat transfer by movement of warm air to cooler part.

Cooling system – Usually metal pipes linked to cold water supply used to remove heat from inside an incubator if the temperature rises above set point.

Copulation – The act of sexual intercourse.

Corrosive – Tending to eat away or consume.

Cuticle – Outermost layer on avian eggshell, usually organic in origin.

Daughter cell – Product of cell division.

Density – Weight per unit of volume.

Dehydration – Lower than normal water content in body.

Detergency – Degree of cleaning power.

Development – Progress of growth and differentiation.

Differentiation – Process of change from simple to more specialised form.

Diffusion – Transfer of a chemical from an area of high concentration to an area of low concentration.

Diploid – Cell containing a set of paired chromosomes.

Disease – Any impairment of normal physiological function affecting an organism.

Disinfection – Rid of potentially harmful micro-organisms.

Distal – Situated furthest from the centre.

DNA – De-oxyribonucleic acid.

Dorsal – Upper surface.

Down – Soft feathers covering chick.

Dry bulb thermometer – Temperature probe exposed to air.

Egg – Reproductive structure.

Egg debris – Pieces of shell and membranes left after hatching.

Egg tooth – A hard structure on the tip of the beak used in hatching by many birds but not by ratites.

Egg white – Albumen.

Ejaculate – The amount of semen produced at one ejaculation.

Ejaculation – Release of semen during copulation.

Elongation – Maximum length of egg divided by maximum width.

Embryo – An bird in the early stages of development up to hatching.

Enzyme – Biological catalyst.

Evaporative cooling – Loss of heat through the evaporation of water vapour from a surface.

Evaporation – Loss of water vapour from the surface of liquid water.

External pipping – Breaking the eggshell during hatching.

Extra-embryonic – Outside of the embryo.

Faecal – Of the faeces.

Fertile – Capable of undergoing growth and development.

Fertilisation – Process of union of spermatozoon and ovum.

Fertility – The state of being fertile.

Fixed-rack – Arrangement where the carriers for egg trays are fixed within the cabinet of the incubator.

Fog – Suspension of liquid droplets in air.

Fogging – Production of fine mist of sanitising chemical.

Folic acid – A B-complex vitamin.

Force-draught – Ventilation of air in incubator using a fan.

Fumigation – To treat with gas or smoke.

Fungi – Plant-like organism lacking chlorophyll, stems and roots and reproducing by spores (*e.g.* mushroom or mould).

Gastrula – Stage of early development where embryo consists of three layers of cells.

GH₂O – Water vapour conductance.

Gradient – Measure of the change in a specific compound or energy between two specific points.

Growth – Increase in size.

Haemoglobin – Respiratory pigment used in red blood cells to transport oxygen.

Haploid – Cell containing a set of single chromosomes.

Hatchability – Number of chicks hatching from eggs.

Hatcher – Machine where eggs hatch.

Hatchery – Building holding incubators and hatchers.

Hatchling – Young bird within 1-2 days of hatching.

Heaters – Suppliers of heat within incubators usually powered by electricity.

Hot spot – Part of incubator that is warmer than rest of machine.

Humidistat – Electronic instrument for controlling humidity.

Humidity – Amount of water vapour in air.

Humidity bottle – Reservoir of water for the wick on a wet bulb thermometer.

Humidity pan – Reservoir of water for producing humidity in an incubator.

Hygiene – Maintenance of health.

Hypercapnia – Excess of carbon dioxide.

Hypocapnia – Lack of carbon dioxide.

Hypoxia – Lack of oxygen.

Incubation – Maintenance of temperature and gaseous environment to allow embryonic development.

Incubation period – The period of time from setting of the egg to hatching of the chick.

Incubator (setter) – Machine in which eggs incubate.

Infertile – Not capable of under-going growth and development.

Infundibulum – Open top of oviduct.

Inorganic – Does not contain carbon as the main element.

Insemination – Introduction of spermatozoa into the reproductive parts of the female.

Internal pipping – Penetration of embryo's beak into air space during hatching process.

Isthmus – Part of oviduct where shell membranes are deposited.

Joule (J) – Unit of energy.

Kidney – Organ which filter out waste chemicals from the blood to produce urine.

Kilogram (kg) – 1,000 grams (2.2 imperial pounds).

Kilojoule (kJ) – 1,000 joules.

Lung – Bodily organ used to gas exchange.

Lysozyme – Enzyme found in egg albumen which destroys bacterial cell walls.

Magnum – Part of oviduct where albumen proteins are deposited.

Malformation – Result of problem during development where normal bodily structures are disrupted and the embryo is malformed.

Malposition – Inappropriate position adopted by embryo which prevents it from hatching.

Maturation – Becoming fully developed.

Medial – To the middle.

Metabolic heat – Energy produced during normal metabolism.

Metabolism – The chemical processes that occur in living organisms.

Microbial – Relating to micro-organisms.

Minerals – Inorganic substances, particularly metallic ions.

Mortality – Death.

Multi-stage incubation – The eggs in the incubator are a variety of different ages.

Natural incubation – Eggs are sat on by the bird.

Navel – Scar on abdomen where embryonic umbilicus joined body.

Neonatal – Newly born.

Nest site – Location where eggs are laid on regular basis.

Nidifugous – Fully feathered and leaving the nest soon after hatching.

Non-invasively – Without entry into the egg by breaking the shell.

Nutrient – Substance which nourishes an animal.

Nutrition – Intake and assimilation of nutrients.

O_2 – Oxygen.

Oedema – Accumulation of fluid in tissue.

Organic – Compound based around carbon as the main element.

Organism – Any living animal, plant or microbe.

Ovary – Female reproductive organ.

Overheating – Temperature rises above normal levels.

Oviduct – Tube down which eggs pass during formation from ovary to cloaca.

Ovomucoid – Protein found in egg albumen.

Ovotransferrin – Protein found in egg albumen which binds iron.

Ovulation – Release of the ovum from the ovary.

Ovum – Female sex cell (= yolk in birds).

Ova – Female sex cells.

Oxygen – A gas forming 21.94% of the air consumed during cell metabolism.

Oxygen consumption – Uptake of oxygen from the air.

Palisade layer – Bulk of calcitic egg-shell.

Pantothenic acid – A B-complex vitamin.

Partial pressure – Fraction of atmospheric pressure created by the presence of one gas.

Pathogen – Agent which causes disease.

Pathology – The study of disease.

P_eCO_2 – Partial pressure of carbon dioxide outside egg.

P_eH_2O – Partial pressure of water vapour outside the egg.

P_eO_2 – Partial pressure of oxygen outside the egg.

pH – A measure of the number of hydrogen ions in solution. Equivalent to acidity.

Physiological zero – Temperature at which no embryonic development occurs – 25°C in birds.

P_iCO_2 – Partial pressure of carbon dioxide inside the egg.

P_iH_2O – Partial pressure of water vapour inside the egg.

P_iO_2 – Partial pressure of oxygen inside egg.

Pip hole – Initial break in the egg-shell.

Pipping – Breaking through into the air space or through the shell.

Pipping muscle – Muscle on nape of neck used to help break shell during hatching.

Plumage – Feathers.

Pore – Small, tube-like opening in eggshell.

Porosity – Measure of the number and depth of pores in an eggshell.

Posterior – To the rear.

Post mortem – After death, usually referring to examination of cause of death.

Poultry – Fowl, turkeys and domestic ducks.

Precocial – Fully feathered and active being able to feed soon after hatching.

Pre-warming – Increasing the temperature of stored eggs prior to setting.

Protein – Chemical based around nitrogen forming part of living organisms.

Proximal – Located close to centre.

Psychrometric chart – Diagrammatic illustration of the relationships between temperature and humidity.

Pull eggs – Continually remove eggs from nest in order to stimulate formation of more.

Radiation – Direct transfer of energy.

Relative humidity – The amount of water vapour in the air as a percentage of the saturation value for water vapour at any one temperature.

Residual yolk – The yolk retracted into the abdomen prior to hatching which sustains a chick during the first few days of life.

Respiration – Process of taking in oxygen and losing carbon dioxide.

Respiratory gas – Oxygen or carbon dioxide.

Respiratory gas exchange – Uptake of oxygen and loss of carbon dioxide.

Respiratory quotient – The ratio of carbon dioxide produced to oxygen consumed during metabolism.

Riboflavin – A B-complex vitamin.

Sanitising – Making hygienic.

Saturation – The maximum amount of water vapour that air can hold.

Scanning Electron microscopy – Process of magnifying objects using electrons rather than light to form an image.

Semen – Fluid containing spermatozoa.

Set point temperature – Incubator temperature at which the air in the cabinet is maintained.

Setter – Incubator.

Setting – Placing in incubator.

Sex determination – The process by which gender is fixed during development.

Sexual reproduction – Procreation by the combining of male and female sex cells.

Shell – Hard outer surface of egg.

Shell gland – Part of oviduct where calcitic shell is deposited.

Shell membranes – Two fibrous layers (inner and outer) between true shell and albumen.

Single stage incubation – All of the eggs in the incubator are the same age.

Solar radiation – Heat from the sun.

Sperm – Male sex cell.

Spermatozoon – Male sex cell.

Spermatozoa – Male sex cells.

Sporicidal – Kills microbial spores.

Sterilise – Remove all life.

Still-air – Without any active air movement.

Storage – Keeping eggs prior to incubation.

Surface crystal layer – Thin layer of calcite crystal on external surface of eggshell.

Synchrony – Occurring at the same time.

Testis – Male sex organ.

Thermal inertia – Degree of retention of heat by a body.

Thermometer - Instrument for measuring temperature.

Thiamin – A B-complex vitamin.

Thorax – Part of body enclosed by ribs.

Toxicity – Degree of harmfulness to people.

Transfer – Time when eggs are moved from the incubator to the hatcher.

Tray – Used to hold eggs during incubation.

Trolley – Carriers for egg trays are located on a moveable trolley which can be taken out of the setter.

Tubule – Small tubular structure.

Turning – Movement and rotation of eggs during incubation.

Unpipped – Unbroken during hatching.

Urates – Salts of uric acid.

Urea – Crystalline by-product of protein metabolism.

Uric acid – Crystalline by-product of protein metabolism

Urine – Excretion by kidneys.

Urodeum - Part of the cloaca of birds.

Ultraviolet (UV) light – Light which is part of electromagnetic spectrum with wavelength shorter than light.

Vapour pressure – see partial pressure.

Vascular – Rich in blood vessels.

Ventilation – Movement and exchange of air within a room or incubator.

Ventral – Lower surface.

Virucidal – Kills viruses.

Viscera – Abdominal organs.

Vitamin – Food item essential for normal nutrition.

Vitelline membrane – Fibrous membrane surrounding the yolk material.

Water vapour – Gaseous water.

Water vapour conductance – Measure of the porosity of eggshells to gaseous water.

Watt – Unit of energy production (joules per second).

Weight loss – Measure of weight loss from egg.

Wet bulb thermometer – Temperature probe covered by a wet cotton wick.

White yolk – Yolk immediately below the blastoderm.

Wick – Cotton layer or tube which fits over a temperature probe and is connected to a water reservoir.

Yellow yolk – Bulk of yolk material.

Yolk – Spherical body in egg supplying nutrients for embryo.

Yolk sac – Membranous structure containing yolk.

Yolk sac membrane – Vascular membrane surrounding yolk.

Zygote – product of the fusion of the female ovum and the male sperm.

Further reading

There are hundreds of scientific papers covering all aspects of incubation but many of these are unavailable, and often incomprehensible, to the general public. As such I feel that it is pointless trying to provide a detailed list of references for the information in this book. Instead, below are several books that review these scientific publications (marked by *) whilst others provide practical information about artificial incubation. To the best of my knowledge all are currently commercially available.

Avian Growth and Development. Evolution within the Altricial-Precocial Spectrum. Edited by J. M. Starck & R. E. Ricklefs (1998). Oxford University Press, Oxford. ISBN 0-19-510608-3.

Avian Incubation. Edited by S. G. Tullett (2002). Butterworths-Heinemann, London. ISBN 0-7506-1002-6

Avian Incubation: Behaviour, Environment and Evolution. Edited by D. C. Deeming (2002). Oxford University Press, Oxford. ISBN 0-19-850810-7.

Bird Nests and Construction Behaviour. Mike Hansell (2001). Cambridge University Press, Cambridge. ISBN 0-521-46038-7.

Egg Incubation: Its Effects on Embryonic Development in Birds and Reptiles. Edited by D. C. Deeming & M. W. J. Ferguson (1991). Cambridge University Press, Cambridge. ISBN 0-521-39071-0.

Egg Production and Incubation. The Game Conservancy (1993). Game Conservancy Trust, Fordingbridge.

Parrot Incubation Procedures. Rick Jordan (1989). Silvio Mattacchione and Co., Ontario. ISBN 0-9692640-7-0.

Practical Incubation, revised and updated edition. Rob Harvey (1993). Hancock House Publishers Ltd., British Colombia. ISBN 0-88839-310-5.

Principles of Artificial Incubation for Game Birds - A Practical Guide. D. C. Deeming (2000). Ratite Conference Books. ISBN 0-952758-3-1.

Ratite Egg Incubation - A Practical Guide. D. C. Deeming (1997). Ratite Conference Books. ISBN 0-952758-2-3.

The New Incubation Book. A. F. Anderson Brown & G. E. S. Robbins (1994). World Pheasant Association. ISBN 0-86230-061-4.

The Ostrich: Biology, Production and Health. Edited by D. C. Deeming. (1999). C.A.B. International, Wallingford. ISBN 0-85199-350-8

Species index

Subject index

Figures in italics indicate an illustration